RENEWABLE CARBON

Advances in Green and Sustainable Chemistry

RENEWABLE CARBON

Science, Technology, and Sustainability

Series Editors

BELA TOROK

TIMOTHY DRANSFIELD

Authored by

SÍLVIO VAZ, JR.

Research Scientist at Brazilian Agricultural Research Corporation, Brasília, Brazil

Professor of the Biofuels Graduate Program, Federal University of Uberlândia, Uberlândia, Brazil

Professor of the Environmental Engineering Graduate Program, Federal University of Ouro Preto, Ouro Preto, Brazil

ELSEVIER

Elsevier
Radarweg 29, PO Box 211, 1000 AE Amsterdam, Netherlands
The Boulevard, Langford Lane, Kidlington, Oxford OX5 1GB, United Kingdom
50 Hampshire Street, 5th Floor, Cambridge, MA 02139, United States

ISBN: 978-0-323-99735-5

Publisher: Candice Janco
Acquisitions Editor: Charlotte Rowley
Editorial Project Manager: Catherine Costello
Production Project Manager: Kiruthika Govindaraju
Cover Designer: Matthew Limbert

Typeset by STRAIVE, India

Dedication

I would like to dedicate this book to my family: my father, Silvio Vaz; my sisters, Ana Maria and Luiza; my wife, Ana Suelen; and my daughter, Elena.

Contents

Part 3 Sustainability

Preface

The Renewable Carbon (RC) concept can be defined as *the use of CO_2, biomass, and recyclable materials as industrial raw materials with an expected reduction in the environmental impact on the related value chains.* From this concept, we can derive several scientific and technological strategies to discover more sustainable industrial products and processes for modern society.

We can salvage RC sources from modern society's products and processes: biomass (e.g., crops, agroindustrial residues, animal residues), carbon dioxide (CO_2) (e.g., generated from the combustion of fossil fuels), and recycling materials (e.g., plastics for industrial and various uses).

The development and production of sustainable materials, fuels, and chemicals is the subject of efforts of applying the RC concept to the real world. And these classes of renewable products should promote positive impacts on economy, society, and the environment.

This book deals with the RC concept proposal for modern society to reduce or substitute demand for fossil feedstocks for chemicals, fuels, and materials. It is divided in three parts comprising aspects of science, technology, and sustainability. From RC sources, a large variety of transformation industries could produce more sustainable products and processes, for example, a circular economy allied to green chemistry principles. The search for renewable chemicals, fuels, and materials is an opportunity to develop and establish environmental-friendly value chains based on biomass, CO_2, and recycling materials.

The author carried out a careful survey of the main industrial technologies in use, using his academic and professional experiences. Moreover, the author also considered trends observed in current research, development, and innovation scenarios; regulatory legislation; the market and other relevant aspects, such as the principles of green chemistry; and the proposal of more sustainable chemistry to reach sustainable value chains.

Seventeen chapters compose the book: Chapter 1 deals with the definition of the RC concept and physicochemical properties of the most representative carbonaceous materials. Biomass, carbon dioxide, and recyclable plastics are potential raw materials for several products, such as chemicals, materials, and pharmaceuticals, among others; and these materials are presented and discussed. The RC strategy can be a valuable contribution to a circular economy because it promotes renewability as the industrial driver for carbonaceous raw materials and helps reduce the greenhouse gas effect and, consequently, climate change. From these observations, RC can contribute to develop sustainable value chains for modern society.

Chapter 2 deals with relevant organic synthetic routes and processes to obtain carbon-based chemicals and materials. Moreover, it includes a case study for a better visualization of the role of organic synthesis in the RC concept.

Chapter 3 deals with the main inorganic and thermochemical routes and processes to obtain carbon-based chemicals and materials. Moreover, it includes two case studies for chemicals (nanocarbonate) and materials (carbon/carbon composite) to reinforce the relevance of both synthetic strategies.

Chapter 4 deals with the main biochemical routes and processes to obtain carbon-based chemicals and fuels by means of microbial and enzymatic conversions. Metabolic and kinetic aspects are also discussed. Moreover, it includes a case study based on succinic acid production.

Chapter 5 deals with the main conclusions and remarks of Part 1 to reinforce the content.

Chapter 6 deals with an introductory text about the main technologies for the RC usage as sustainable industrial feedstock. Sectors such as agriculture and agroindustry, energy generation, chemical industry, iron, steel, cement, plastics (in a general way), resins, and synthetic fibers are discussed. Furthermore, the technology readiness level (TRL) assessment is presented and discussed as it relates to those technologies for RC use.

Chapter 7 deals with technological strategies for carbon capture use and carbon capture use and storage (CCU/CCUS) such as adsorption, absorption, membrane separation, and chemical capture. Several process diagrams of these technologies are presented in detail. Moreover, the chapter includes a case study using the TRL scale for technology assessment.

Chapter 8 deals with the definition of nanotechnology and its related technologies and products for the three RC raw material sources. Analytical techniques for monitoring and control are also presented. Moreover, it includes case studies related to the production of absorbent, nanotubes, and flash graphene to demonstrate these technological and economic potentials.

Chapter 9 deals with biotechnological aspects, technologies, and products for biomass as the main source of RC. The current scenario associated with green chemistry and analytical technologies are also discussed. Moreover, it includes a case study based on the use of agroindustrial waste.

Chapter 10 deals with thermochemical aspects, technologies, and products for agroindustrial biomass and recycling materials as the main sources of RC. Basic analytical techniques are also presented. Moreover, it includes case studies for each RC renewable material.

Chapter 11 deals with homogeneous and heterogeneous catalysis, their technologies, and products for carbon dioxide and agroindustrial biomass as the main sources of RC. Status of technologies and analytical techniques are also discussed. Moreover, it includes several case studies involving heterogeneous and homogeneous catalysts.

Chapter 12 deals with the main conclusions and remarks of Part 2 to reinforce the content.

Chapter 13 deals with the definition of the sustainability concept. Moreover, Sustainable Development Goals, circular economy, and green chemistry are presented and explored for RC. Aspects of climate change and the influence of biogenic carbon and anthropogenic carbon on this phenomenon are also discussed. To finalize, aspects of pricing and marketing are also considered.

Chapter 14 deals with the main metrics applied to determine the sustainability of products and processes related to RC, considering the lifecycle assessment (LCA) as the main methodology. Moreover, it includes a case study related to the biofuel sector.

Chapter 15 deals with the construction of new sustainable value chains based on RC as a substitution for petrochemicals. Sustainable strategies for these value chains are also discussed. Moreover, it includes a case study related to building blocks from biomass for renewable plastics.

Chapter 16 deals with the main strategies to establish ecofriendly processing approaches for RC by means of the industrial ecology. Moreover, it includes a case study related to the chemical industry.

Finally, Chapter 17 deals with the main conclusions and remarks of Part 3 to reinforce the content.

Thus, the book presents a review of sustainable feedstocks containing RC, i.e., biomass, CO_2, and recycling materials; a compendium of technologies to explore RC sources; and a compendium of renewable products and materials to be obtained by the industry. Moreover, the book offers information to the reader based on start-of-art and practical applicability of its content.

Good lecture!

Sílvio Vaz, Jr.

PART 1

Science

CHAPTER 1

Introduction to renewable carbon—Concept and properties

The definition and application of renewable carbon (RC) is comprised of several aspects, from environmental chemistry, to biotechnology, nanotechnology, materials science, and chemical engineering, among others, transforming it into a very promising and exciting theme, which will be explored in this introductory chapter.

First, it is important to review some of the main physicochemical properties of the carbon atom (from the Latin *carbo*: charcoal):

✓ Atomic number (Z): 6
✓ Periodic Table: group 14 and period 2
✓ Electronic configuration: $[He]2s^2 2p^2$
✓ Atomic radius: 1.70 Å (nonbonded)
✓ Covalent radius: 0.75 Å
✓ Key isotopes: ^{12}C (atomic mass 12.0107), ^{13}C (atomic mass 13.003), ^{14}C (atomic 14.003)
✓ Physical state: solid state (according the STPC[a])
✓ CAS number: 7440-44-0
✓ Density: $3.513\,g\,cm^{-3}$ (diamond), and $2.2\,g\,cm^{-3}$ (graphite)
✓ Allotropes: diamond, graphite, graphene, amorphous, and fullerene
✓ Boiling point 4027°C
✓ Melting point 3527°C

Regarding its presence, carbon is found on Earth in rock minerals, coal, oil, and natural gas, in addition to being present in all living beings (biomass). Furthermore, we can observe various types of carbon as a solid, as amorphous carbon, carbon fibers, carbon materials, glass-like carbon, graphitic carbon, nongraphitic carbon, and pyrolytic carbon.

Carbon has different allotropic forms, with graphite and diamond being the most important. Graphite (from the Greek *gráphos*: to write) has been

[a] STPC = Standard Temperature and Pressure Conditions, https://www.sciencedirect.com/topics/engineering/standard-temperature-and-pressure.

Renewable Carbon
https://doi.org/10.1016/B978-0-323-99735-5.00007-4

known and used since antiquity in the manufacture of pencils. It is a crystalline substance, gray in color, found in nature, and can be obtained through mining and purified through the flotation process. It can also be produced artificially by reacting fossil coal, coke, and quartz at high temperatures. Diamond (from the Greek *adámas*: invincible) is colorless, extremely hard, and the only gemstone composed of a single element. It is found in nature associated with volcanic rocks, such as kimberlite, and in deposits. It can be produced artificially by heating graphite under high pressures and in the presence of a catalyst. The most important carbonaceous minerals are dolomite magnesite, siderite and zinc, and sodium and calcium carbonates.

Carbon can form a huge number of compounds, the most important are the carbonaceous substances, such as:

- Carbon oxides: monoxide (CO) and dioxide (CO_2)
- Carbon disulfide (CS_2)
- Hydrocarbons: combinations of carbon and hydrogen found in oil and natural gas
- Compounds formed by salts of carbonic acid, such as carbonates and bicarbonates

Furthermore, it is possible to obtain carbon by burning charcoal from wood or other lignocellulosic residues.

It is important to emphasize that the biogeochemical carbon cycle (Fig. 1.1) is essential for life and the ecological balance of the environment.

Indeed, the natural concentration of carbon dioxide in the Earth's atmosphere was 280 ppm around 150 years ago. In 2013, as a result of combusting fossil fuels with oxygen, there was 390 ppm (Royal Society of Chemistry, 2021). Atmospheric carbon dioxide allows in visible light but prevents some infrared light from escaping (the natural greenhouse effect). This keeps the Earth warm enough to sustain life. However, an enhanced greenhouse effect is underway due to a human-induced rise in atmospheric carbon dioxide, and it's affecting living things by means climate changes.

As for the applications of carbonaceous substances, prehistoric man already used carbon in the form of charcoal. Currently, there are a large number of diverse applications, such as:

✓ Charcoal, known as activated charcoal after treatment, is used to purify water, remove odors and radioactive residues, recover solvents, manufacture gas masks, and produce medicine.

✓ The charcoal known as carbon black is used in the manufacture of printing inks, greases, and black enamels.

✓ Various types of coal (vegetable and mineral) are used as fuels, lubricants, and propellants in thermal power stations.

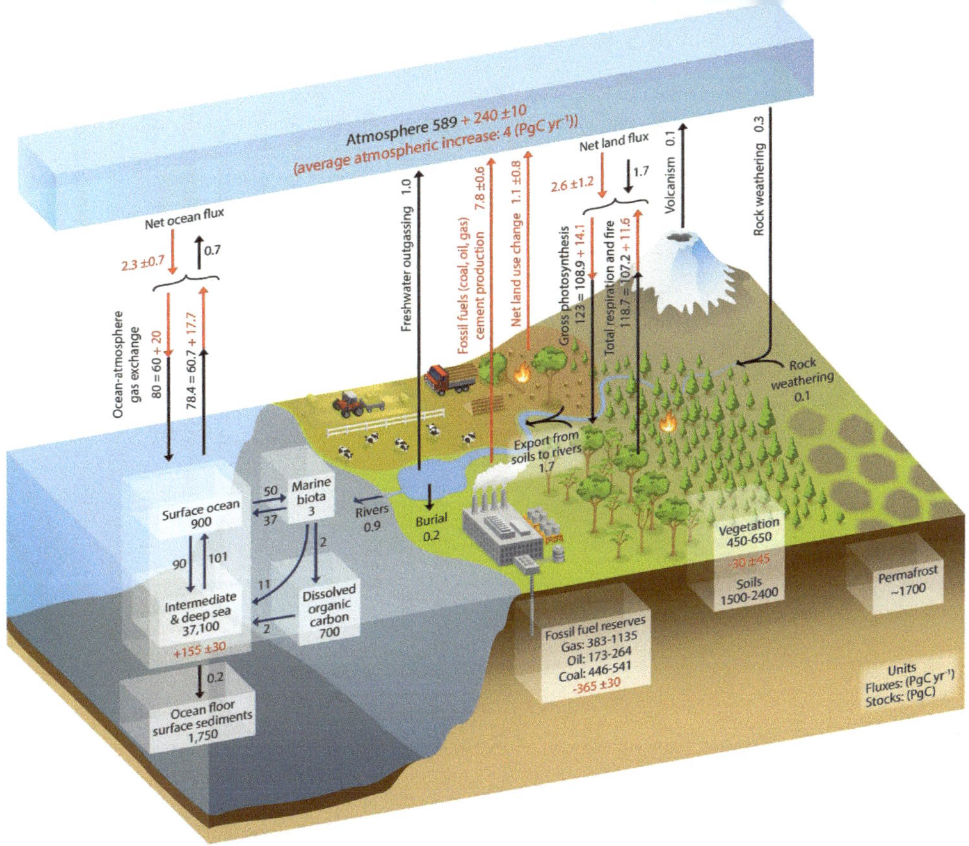

Fig. 1.1 A simplified schematic of the global carbon cycle. *Numbers* represent reservoir mass (i.e., carbon stocks) in petagrams of carbon (PgC) and annual carbon exchange fluxes in PgC $year^{-1}$. *Black numbers* and *arrows* indicate reservoir mass and exchange fluxes estimated for the time prior to the Industrial Era, about 1750. *(Credit: Ciais, P., Sabine, C., Bala, G., Bopp, L., Brovkin, V., Canadell, J., Chhabra, A., DeFries, R., Galloway, J., Heimann, M., Jones, C., Le Quere, C., Myneni, R.B., Piao, S., Thornton, P., 2013. Carbon and other biogeochemical cycles. In: Stocker, T.F., Qin, D., Plattner, G.-K., Tignor, M., Allen, S.K., Boschung, J., Nauels, A., Xia, Y., Bex, V., Midgley, P.M. (Eds.), Climate Change 2013: The Physical Science Basis. Contribution of Working Group I to the Fifth Assessment Report of the Intergovernmental Panel on Climate Change. Cambridge University Press, Cambridge. Reprinted with permission from Intergovernmental Panel on Climate Change.)*

- ✓ Carbon fibers are used to improve the mechanical properties of other fibers, especially textiles.
- ✓ In the steel industry, carbon is used in foundries in the form of coke and charcoal as a reducer to obtain iron in the blast furnace and produce steel.
- ✓ Diamond can be used in industry and in cutting tools to cut other softer pieces.
- ✓ Graphite is used in the manufacture of pencils, solid lubricants, electric brushes, batteries, rubbers, coatings, and in electroplating.
- ✓ Pyrographite, a fibrous form of graphite, can be used as part of the fuselage of rockets, missiles, and other space vehicles.
- ✓ Bicarbonates are used in the production of antiacids.
- ✓ Precursors for synthesis of chemicals, fuels, medicines, plastics, resins, fibers, etc.

Thus, carbon science, technology, and use are very broad, involving different economic sectors and environmental issues, as we will see in detail ahead.

1.1 Introduction

Renewability is a condition whereby a certain material or molecule, i.e., precursor or raw material, can be used in a circular pathway without its depletion.

The circular economy, an economic strategy to reach renewability, is defined by the European Parliament (2021) as *"a model of production and consumption that involves sharing, leasing, reusing, repairing, reforming, and recycling existing materials and products for as long as possible. In this way, the lifecycle of products is extended"* (see Part 3). In practice, it means reducing waste to a minimum. When a product reaches the end of its useful life, its materials are kept in the economy whenever possible and they can be productively used repeatedly, thus creating more value. For example, the use of a circular economy for biomass residues—one of the main sources of carbonaceous materials—may involve (Vaz Jr., 2020):

(i) Reduction in the generation of residues through the use of all by-products and/or co-products generated during processing
(ii) The creation of new value chains based on the use of waste

According to the Organisation for Economic Co-operation and Development (2018), a "*bioeconomy refers to the set of economic activities related to the invention, development, production, and use of biological products and processes.*" That said, it is easy to see that the bioeconomy has a broad scope, and its applicability to biomass as source of RC is obvious. Furthermore, the bioeconomy is in line with the aforementioned United Nations sustainable development goals (United Nations, 2021) due to the promotion of sustainable value chains based on renewable resources. New opportunities arising from the bioeconomy model are widely found in manufacturing, biochemistry, and agriculture; however, exploration strategies must also include accelerated innovations for food security and the protection of natural resources (von Braun, 2018).

From these statements, the *renewable carbon strategy* based on the reuse of the carbon sources can be a valuable contribution to rethink the established economic model to promote the circular economy and the bioeconomy, because it considers carbonaceous raw material biomass, carbon dioxide, and recyclable materials as industrial sources (Renewable Carbon Initiative, 2021).

1.2 The renewable carbon concept in modern society

The RC concept (Fig. 1.2) can be defined as *the use of CO_2, biomass, and recyclable materials as industrial raw materials with an expected reduction in the*

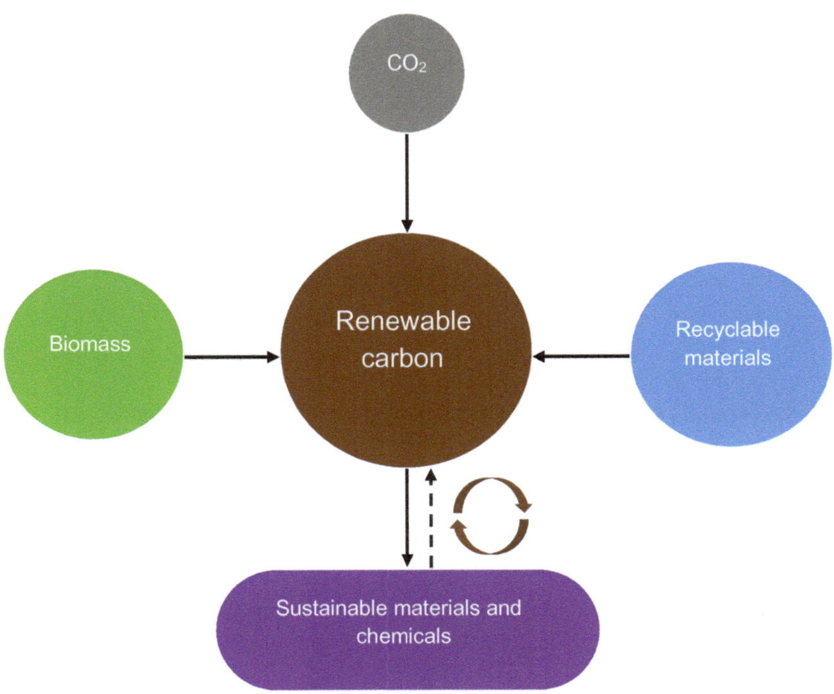

Fig. 1.2 An illustration of the renewable carbon concept, considering a cycling pathway for raw materials and products. *(Credit: author.)*

environmental impact on the related value chains. From this concept, we can derive several scientific and technological strategies to reach more sustainable industrial products and processes for modern society.

We can consider RC sources from products and processes from modern society:

- Biomass: crops, agroindustrial residues, animal residues. The first and second items are the most common for industrial purposes to generate derived products, such as food, feed, biofuels, polymers, materials, etc.; however, animal proteins and fats are used also for feed and biofuel, respectively.
- Carbon dioxide (CO_2): generated from the combustion of fossil fuels (i.e., coal, natural gas, and diesel oil) for power generation (electricity), combustion of fossil fuels (e.g., gasoline and diesel oil) for transportation, cement industry, biomass combustion from agricultural activities and biogas generation, among others.
- Recycling materials: plastics for industrial and various uses and papers from packages and common uses.

Those examples describe representative RC sources based on carbonaceous materials. In some cases, such as for packages, plastics, and papers, these materials are associated to metals, which means inorganic components are also expected to be treated as contaminants.

The development and production of sustainable materials and chemicals is the subject of efforts of the RC concept applied to the real world; both classes of renewable products should promote positive impacts on economy, society, and the environment, as discussed in Part 3 of this book.

Fig. 1.2 also illustrates a smart concept to reach the circular economy by means the renewability of raw materials and their products.

1.3 Introduction to carbonaceous raw materials for industrial purposes

According to the previous definition of RC, we can consider carbonaceous materials as those derived from biomass, carbon dioxide, and recyclable materials used as sources of raw materials for industrial purposes. Fig. 1.3 depicts a carbon fiber obtained from lignin, the macromolecule constituent of plants.

Fig. 1.3 A carbon fiber sample obtained from lignin macromolecule processing. *(Credit: author.)*

Biomass

The need to develop renewable raw materials for industrial chemistry as a substitute for oil has been a strategic challenge for the 21st century. In this context, the use of different types of plant biomass—highlighting starchy, lignocellulosic, oleaginous, and saccharide—can be consolidated as both an alternative of using cheaper and less polluting raw materials, as well as a model of aggregation of economic value to the agroindustrial chains, such as soybeans, sugarcane, corn, and forests, among others. These lines of action may, above all, contribute to the sustainability of a wide range of chemicals, especially organic chemicals (e.g., organic acids, esters, alcohols, sugars, phenolics, etc.), which are widely used in modern society for feed, food, pharmaceuticals, beauty and care, and agrochemicals, among others.

The high heterogeneity and consequent large chemical complexity of the plant biomass (Fig. 1.4) makes it a good raw material for various end products, such as energy, food, chemicals, pharmaceuticals, and materials. We can classify plant biomass of great economic interest into four types: oil crop or oleaginous, saccharides (or sugary), starchy, and lignocellulosic. Soybean (*Glycine max*) and palm oil (*Elaeis guinensis*) are examples of oil plant species; sugarcane (*Saccharum* spp.) and sorghum (*Sorghum bicolor* (L.) Moench) are saccharide biomasses; maize or corn (*Zea mays*) is a starchy biomass; and bagasse, straw, and wood are lignocellulosic biomasses.

According to The United Nations Food and Agriculture Organization (2021), the global average production of agroindustrial biomass—as the most produced commodities—is:

- Cereals: 2.4 Gt
- Sugar crops: 1.8 Gt

Fig. 1.4 Examples of agroindustrial biomass: (A) sugarcane bagasse, a lignocellulosic biomass; (B) corn, a starchy biomass; (C) sugarcane, a saccharide biomass; and (D) soybean, an oleaginous biomass. *(Credit: author.)*

- Maize or corn: 0.8 Gt
- Oil crops: 0.7 Gt

This high production can be seen as a huge source of RC for industrial purposes arising from both its components (carbohydrates, proteins, oil, etc.) and the waste from its industrial processing.

Carbon dioxide (CO_2)

Carbon dioxide is a gas originating from several processes, both natural (biogenic) and anthropic. It is produced during breathing by all animals, by fungi and microorganisms that depend directly or indirectly on living or decomposing plants for food, and from anthropic activities, mainly combustion processes using fossil fuels with deleterious effects on the environment. However, carbon dioxide is considered a gas with the potential to be used as a raw material (Poliakoff et al., 2015). As for its commercial applications, it can be used as a solvent for extraction by supercritical fluid, vasodilating agent, anesthetic, antagonist, gas for food packaging, food propellant, and gas for soda; in addition, it is a metabolite of *Saccharomyces cerevisiae* yeast activity used in ethanol production.

According to the International Energy Agency (2021), despite the decline in 2020, global energy-related CO_2 emissions (Fig. 1.5) remained at 31.5 Gt, which contributed to CO_2 reaching its highest average annual concentration in the atmosphere of 412.5 ppm in 2020—around 50% higher than when the industrial revolution began.

Despite the environmental risk from this gas production—with a direct influence on global warming—it can be seen as a promising source of RC if it is considered under scientific and technological criteria. A technological use of CO_2 can be observed in Fig. 1.6.

With industrial growth, the emissions of gaseous pollutants have gained strength. According to the Intergovernmental Panel on Climate Change (Intergovernmental Panel on Climate Change, 2021), approximately 120 Gt of CO_2 per year could be emitted through the year 2100, which will contribute to an increase in the temperature of the planet up to 3.5°C. Because of this, global warming, caused by uncontrolled emissions of gases that promote the greenhouse effect (e.g., carbon dioxide, methane, and

Fig. 1.5 A thermoelectric plant for energy generation. This plant is located in Candiota, Rio Grande do Sul, Brazil. *(Credit: author.)*

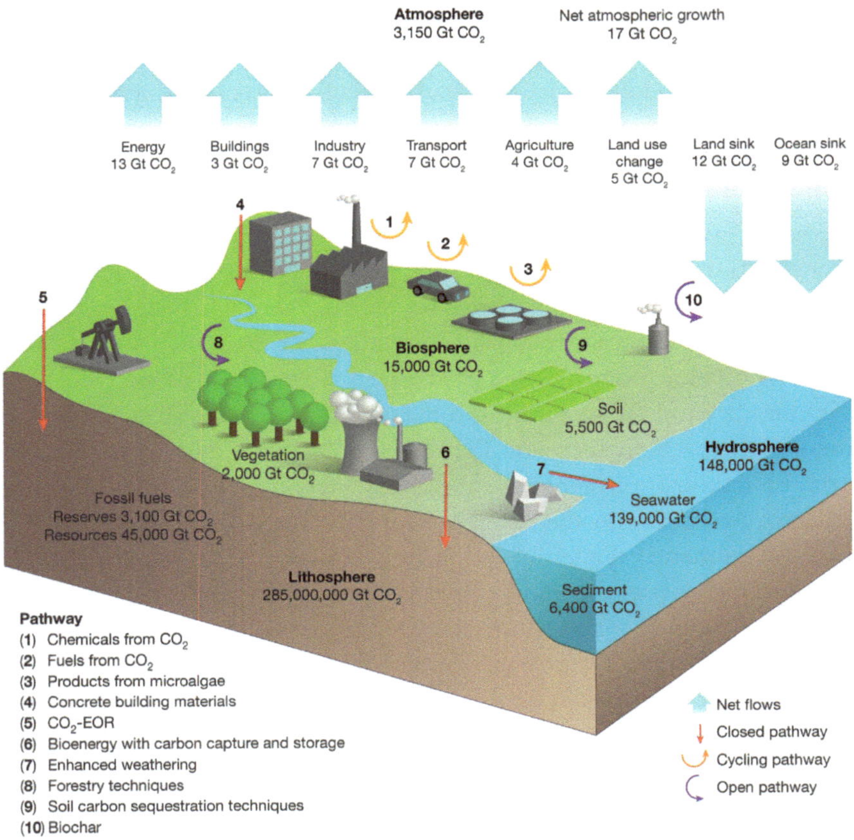

Fig. 1.6 Stocks and net flows of CO_2 including potential utilization and removal pathways, according to Hepburn et al. (2019). *EOR*, enhanced oil recovery. *(Reproduced with permission from Springer Nature.)*

nitrous oxide), is the subject of scientific research, environmental policies, and awareness of modern society, as observed by Vaz Jr. et al. (2022).

The greenhouse effect is a phenomenon that occurs naturally in the Earth's atmosphere. It is a fundamental phenomenon for the existence of life on the planet, as it acts as a blanket receiving solar energy to maintain the average global temperature (Kweku et al., 2017). The greenhouse effect prevents the heat that returns to the atmosphere from being lost to space. Without this condition, Earth would have a temperature so low that there would be no suitable conditions for terrestrial life. In turn, global warming occurs due to the massive presence of gases in the atmosphere that retain heat beyond what is necessary for the maintenance of the Earth (Kweku et al., 2017).

Greenhouse gases (GHG), mostly carbon dioxide, nitrous oxide, and methane, have a molecular structure capable of absorbing a certain amount of heat, and this capacity configures the effect of global warming. If the concentration of molecules in these gases is high, Earth's protective blanket is overloaded with heat, causing global warming. Carbon dioxide contributes 53% to the greenhouse effect (Falci, 2019). Since global warming is a matter of great economic, social, and environmental interest, most countries have directed major actions toward mitigating GHG through the technologies of carbon capture and use (CCU) of the carbon dioxide molecule.

Recyclable materials

The most common source of RC is a material used in a certain value chain, such as plastics and tires.

According to the Environmental Protection Agency (2021), some of the common consumer goods that can be considered recyclables are:

- Paper: makes up 23% of municipal solid waste (trash) generated each year, more than any other material.
- Plastics: more than 35 million tons of plastics were generated in the United States in 2018, which was about 12% of the waste stream; examples of recyclable plastics are polystyrene (PS), polypropylene (PP), low density polyethylene (PE-LD), poly(vinyl chloride) (PVC), high density polyethylene (PE-HE), and poly(ethylene terephthalate) (PET).
- Used motor oil from engines: the used oil from one oil change can pollute one million gallons of fresh water.
- Household hazardous waste: products such as paints, cleaners, oils, batteries, and pesticides that contain potentially hazardous ingredients require special care and may be dangerous to people or bad for the environment.
- Tires.

Considering plastics, the main source of recyclable carbonaceous materials, every year, more than 380 million tons of plastic are produced worldwide. Just 16% of plastic waste is recycled to make new plastics, while 40% is sent to landfills, 25% to incineration, and 19% is dumped (Latham, 2021).

Regarding PET (Fig. 1.7), its characteristics are related to its transparency and resistance to wear and corrosion. It has a high resistance and smooth finish, which explains its wide use in mineral water bottles, soft drinks, juices, edible oils, pharmaceutical products, etc. The material also has low water absorption and good resistance to tensile forces; its strength is similar to that

Fig. 1.7 Recyclable PET from domestic use. *(Credit: Aykut Erdogdu, www.shutterstock.com.)*

of aluminum film. Polyester is generally transparent, and its tensile strength can reach one-third to one-half that of steel if processed by oriented design.

With the large-scale use of PET bottles, especially from the 1990s onward, a serious environmental problem arose: many of these bottles were incorrectly discarded and ended up in landfills, rivers, sewers, seas, and forests. But this material can take up to 800 years to decompose (hence the importance of its collection and recycling). To get rid of the bulk of the bottles, some people may burn the material. When PET is burned, it has a yellowish flame that can explode during burning. But, in addition to being very flammable, the material releases toxins and gases, such as carbon monoxide and carbon dioxide, acetaldehyde, vinyl benzoate, and benzoic acid into the atmosphere, contributing to air pollution (Sovová et al., 2008).

Plastic in general—including PET—are the pollutants most commonly found in the oceans, sometimes as microplastics, an emergent pollutant of environmental concern (Vaz Jr., 2018). In some rotating marine current systems, the amount of plastic found is so large that it has already become part of the ocean. Furthermore, microplastics—plastic pieces with a size less than 5 mm—can be observed in soil, surface water, and groundwater (Wu et al., 2019).

1.4 Physicochemical properties of relevant carbonaceous materials

Biomass

As previously mentioned, here we consider the main classes of agroindustrial biomass, i.e., oleaginous, starchy, saccharide, and lignocellulosic, according definitions established by Vaz Jr. (2016).

Oleaginous biomass is one that has superior fatty acids (Fig. 1.8), and their esters are stored inside seeds or grains, with fatty acids with chains of different sizes and number of unsaturation (lipids). Table 1.1 depicts the fatty acid content for some oleaginous crops.

Oleaginous plants are used as feedstock for the following agroindustrial products:

- Cooking oil obtained from extraction
- Raw material for the agrochemical and chemical industries
- Raw material for biodiesel

Saccharide biomass is one in which the main source of sugar is sucrose, which is a disaccharide formed by glucose and fructose (Fig. 1.9), the latter being its constituent monosaccharides, which are hexoses (C6). Table 1.2 describes the composition of some crops.

Saccharides or sugary plants are used as feedstock for the following agroindustrial products:

- Sugar (saccharose)
- Ethanol
- Raw material for chemical industries
- Raw material for renewable polymers

Starchy biomass has this name because its main chemical constituent is starch, which is a polymer—polysaccharide—whose monomer is glucose (Fig. 1.10), which is a hexose, or six-carbon sugar (C6), a monosaccharide. Table 1.3 depicts the chemical composition of some starchy crops.

As sugary plants, starchy plants are used as feedstock for the following agroindustrial products:

- Sugar (glucose)
- Ethanol
- Raw material for the chemical industries
- Raw material for renewable polymers

Lignocellulosic biomass is the most abundant compared to the others because it is formed by cellulose, hemicellulose, and lignin, which are the three components of the cell wall and the morphological structure of plants. Cellulose (Fig. 1.11) and hemicellulose (Fig. 1.12) are polysaccharide polymers, and lignin (Fig. 1.13) is a phenolic macromolecule. Table 1.4 depicts the contents of lignocellulosic fractions for some crops.

Lignocellulosic plants are used as feedstock for the following agroindustrial products:

- Energy by combustion (co-generation)
- Raw material for renewable polymers

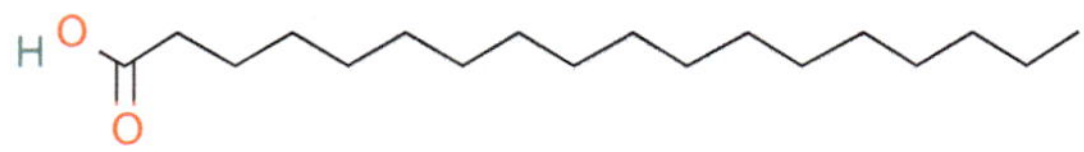

Stearic acid

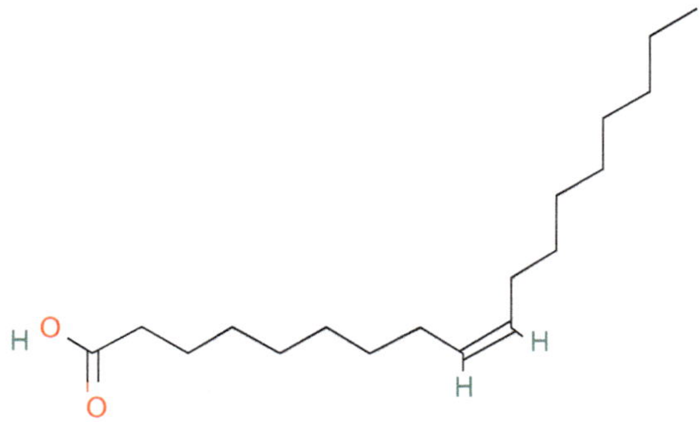

Oleic acid

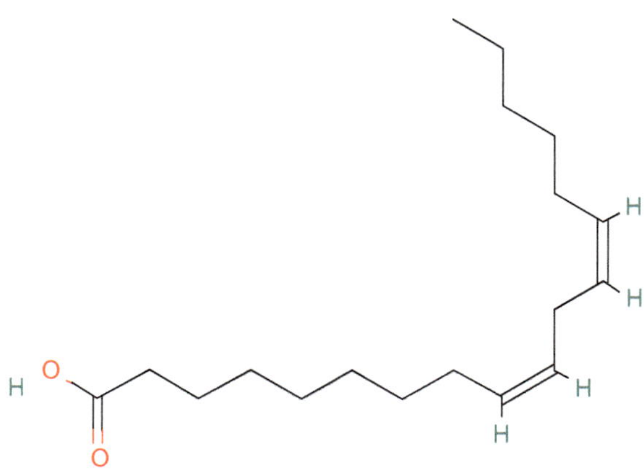

Linoleic acid

Fig. 1.8 Some chemical structures of fatty acids from oleaginous plants such as soybean. *(Credit: author.)*

Table 1.1 Fatty acid composition (% wt/wt) of oils extracted from oleaginous biomass, according to Chen et al., 2020.

	Oil type			
FAC (%)	**Peanut oil**	**Rapeseed oil**	**Soybean oil**	**Linseed oil**
C16:0	10.80 ± 0.01	4.04 ± 0.01	10.14 ± 0.01	5.76 ± 0.01
C18:0	4.18 ± 0.02	1.76 ± 0.00	4.32 ± 0.00	4.22 ± 0.01
C18:1 ω9	41.64 ± 0.02	61.44 ± 0.07	22.62 ± 0.00	18.61 ± 0.01
C18:2 ω6	35.78 ± 0.01	19.30 ± 0.01	52.16 ± 0.04	16.07 ± 0.01
C18:3 ω3	–	7.63 ± 0.01	6.59 ± 0.01	50.61 ± 0.01
C20:0	1.62 ± 0.03	0.61 ± 0.00	0.41 ± 0.01	0.19 ± 0.00
C20:1 ω7	1.03 ± 0.01	1.37 ± 0.02	0.38 ± 0.01	0.12 ± 0.02
C22:0	2.91 ± 0.01	–	–	0.11 ± 0.01
C24:0	1.42 ± 0.00	–	–	0.08 ± 0.00

Results are means ± SD of fatty acid composition (FAC).
Reproduced with permission from Elsevier.

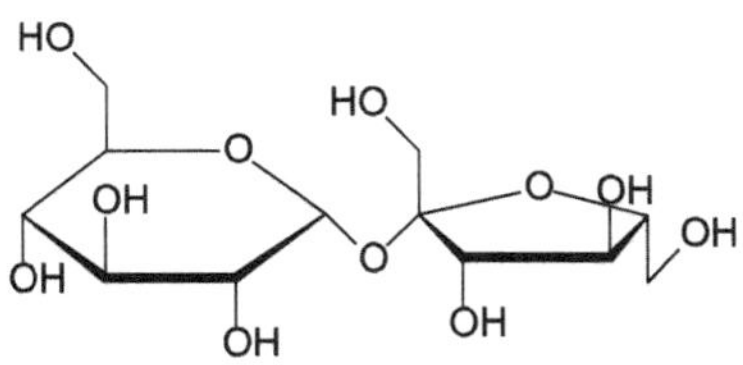

Fig. 1.9 Chemical structure of sucrose, a disaccharide present in sugarcane. The D-glucose moiety on the left and the D-fructose moiety on the right are linked by α-β-D-disaccharide bonds. *(Credit: author.)*

Table 1.2 Chemical composition of broth extracted from sugarcane, according to Faria et al. (2011), and sweet sorghum, according to Mamma et al. (1995).

Crop	% wt/wt sucrose	% wt/wt glucose	% wt/wt organic acid
Sugarcane	85.3	–	24
Sweet sorghum	14.8	1.5	–

Reproduced with permission from Elsevier.

- Cellulose and their derivatives
- Lignin and their derivatives
- Sugars from cellulose (glucose) and hemicellulose (xylose)

It is the main source of agricultural residue due to the fact that all plants contain lignin, cellulose, and hemicellulose. Certainly, it is also the main source of RC from biomass.

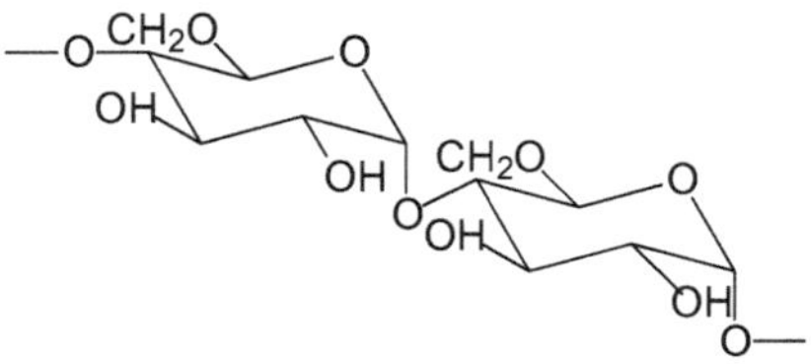

Fig. 1.10 Chemical structure of starch polymer; the glucose unities (monomers) are linked by α-1-4-D-disaccharide bonds. *(Credit: author.)*

Table 1.3 Chemical composition of corn grain flour (Sandhu et al., 2007), cassava (Charles et al., 2005), and potato (Liu et al., 2007).

Crop	% wt/wt starch	% wt/wt protein	% wt/wt fiber	% wt/wt others
Corn (flour of grain)	90.1	6.5	0.52	1.99 (lipd)
Cassava (pulp)	83.8	1.5	2.5	0.2 (lipd)
Potato (pulp)	71.5	8.6	5.4	–

Reproduced with permission from Elsevier.

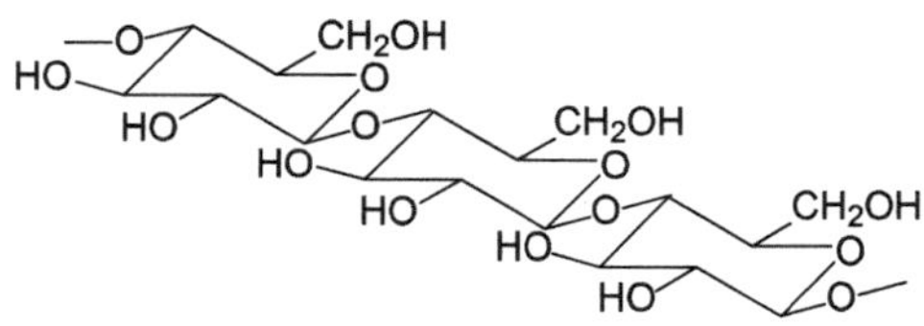

Fig. 1.11 Chemical structure of cellulose; the glucose units are linked by a 1,4-β-D bond. *(Credit: author.)*

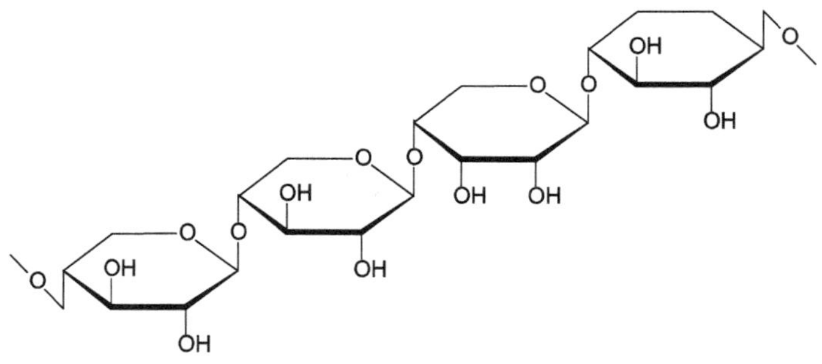

Fig. 1.12 Chemical structure of hemicellulose; the oligomeric units composed of D-glucose and pentoses (mainly D-xylose) are linked by means of a 1,4-β-D bond. *(Credit: author.)*

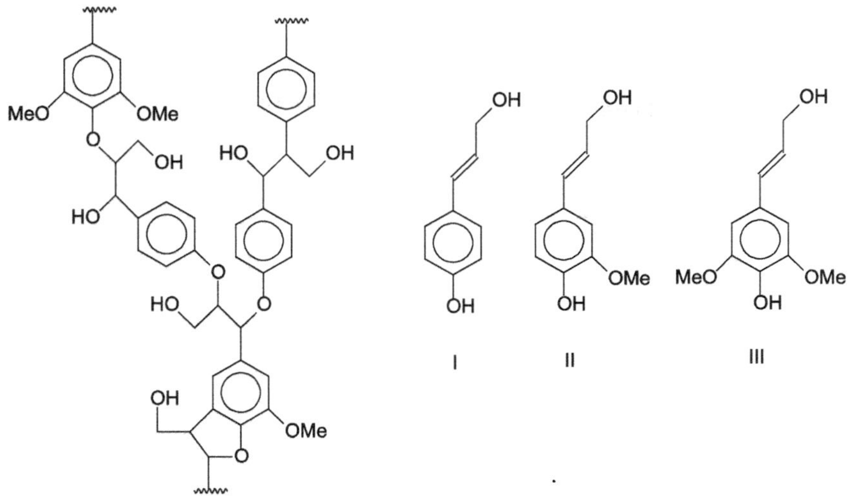

Fig. 1.13 Lignin structure *(left)* and its precursors *(right)*: (I) *p*-coumaryl alcohol, (II) coniferyl alcohol, and (III) sinapyl alcohol. *(Credit: author.)*

Table 1.4 Chemical composition of lignocellulosic biomasses, according to Vassilev et al. (2012).

Crops	% wt/wt cellulose	% wt/wt hemicellulose	% wt/wt lignin
Barley straw	48.6	29.7	21.7
Corn cobs	48.1	37.2	14.7
Grasses	34.2	44.7	21.1
Sugarcane bagasse	42.7	33.1	24.2
Rice husks	43.8	31.6	24.6
Wheat straw	44.5	33.2	22.3
Eucalyptus	52.7	15.4	31.9

Reproduced with permission from Elsevier.

Soil organic matter (OM) originates from the decomposition of plant biomass residues and animal remains that, through chemical, physical, and biological processes, undergo structural modification. This gives rise to a series of organic compounds, whose main representatives are humic substances (HS), which fundamentally contribute to the properties of the soil, as in the case of the capacity of sequestration, mobilization, and redox effect of organic molecules foreign to the environment (xenobiotics) (Stevenson, 1994). The HS interacts with several organic pollutants, such as pesticides and petroleum derivatives, and inorganic ones such as toxic metals (Clapp et al., 2001).

HS is, without doubt, the most abundant source of organic components in nature and are present in all natural soils and waters that contain OM (Stevenson, 1994). It is currently accepted that peat and soil OM are the main active sources of carbon in terms of reversibility of entry and exit of this element in the biogeochemical cycle of nature (Jenkinson et al., 1991).

The constitution of HS, operationally based on solubility in acidic or alkaline media, is as follows (Stevenson, 1994):

- Fulvic acids (FA): fraction soluble under all pH values
- Humic acids (HA): fraction soluble in alkaline medium
- Humine: insoluble fraction under any pH values

Of these three fractions, the HA are more susceptible to structural changes resulting from soil management practices and OM degradation processes (Tatzber et al., 2008).

Methane generated from the biochemical degradation of biomass is a source of additional RC. The anaerobic digestion process involves the degradation and stabilization of OM leading to the formation of methane (55%–70% v/v), inorganic products (mainly carbon dioxide, 30%–45% v/v), and liquid residue rich in minerals that can be used as biofertilizer (stabilized OM). Methane can be used as a fuel (biogas) and for hydrogen production by the reform process (Nalbant and Colpan, 2020). Methane formation is illustrated in Eq. (1.1):

$$\begin{aligned} OM + H_2(g) \rightarrow CH_4(g) + CO_2(g) + NH_3(g) + H_2S(g) + N_2 + CO \\ + O_2 + \text{biomass} + \text{heat} \end{aligned} \tag{1.1}$$

Carbon dioxide (CO_2)

The main physicochemical properties of this gas (Fig. 1.14) are depicted in Table 1.5. It is convenient highlight that the CO_2 molecule is thermodynamically stable.

Recyclable materials

There is a focus on using plastics as the main source of recyclable materials for industrial purposes. Some polymer structures are depicted in Fig. 1.15. Moreover, Table 1.6 describes some physicochemical properties of these plastics.

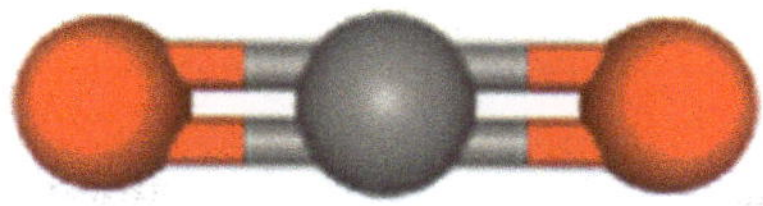

Fig. 1.14 Chemical structure of the carbon dioxide molecule. Oxygen is shown in *red* and carbon in *gray*. *(Credit: Pubchem, 2022. Carbon Dioxide. https://pubchem.ncbi.nlm.nih.gov/compound/280 (accessed March 2022). Reproduced with permission from the U.S. Government.)*

Table 1.5 Physicochemical properties of the carbon dioxide molecule, according to Pubchem (2022).

Property name	**Property value**
Molecular mass	44.009 $g\,mol^{-1}$
Hydrogen bond acceptors	2
Exact molecular weight	43.989829 $g\,mol^{-1}$
Monoisotopic molecular mass	43.989829 $g\,mol^{-1}$
Polar topological surface area	34.1 $Å^2$
Formal cargo	0
Covalently bonded unit	1

Reproduced with permission from the U.S. Government.

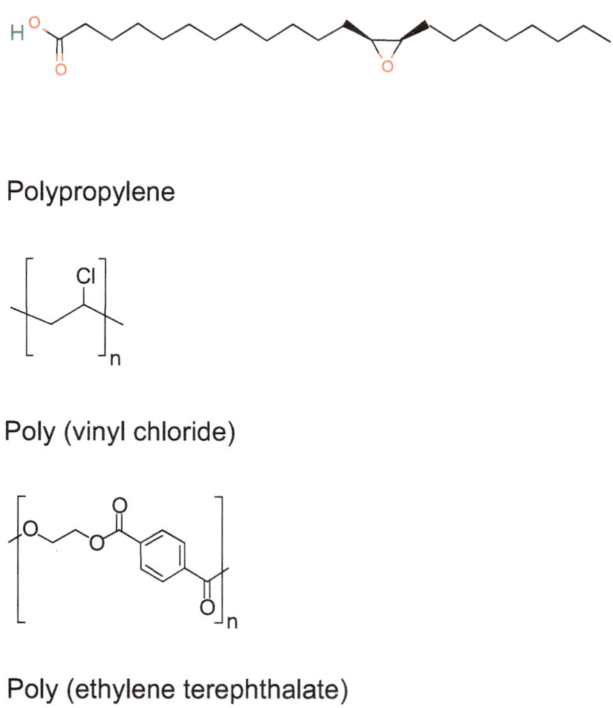

Fig. 1.15 Molecular structures of some commonly used plastics. *(Credit: author.)*

Table 1.6 Physicochemical properties of commonly used plastics.

Plastic	CAS number	Formula	Melting point (°C)	Density ($g\,cm^{-3}$)
Polypropylene (PP)	9003-07-0	$C_{22}H_{42}O_3$	210–290	0.905
Polyethylene (PET)	25038-59-9	$(C_{10}H_8O_4)n$	260	1.37
Poly(vinyl chloride) (PVC)	9002-86-2	$(C_2H_3Cl)n$	92	1.38

The chemical constitution and the physicochemical properties turn the carbon atom from renewable sources a versatile raw material, especialy for organic molecules (as we will see in the Chapter 2). It is important to clarify that new value chains from RC can involve:

✓ New products, as new molecules are developed from RC sources, e.g., from biomass
✓ New processes, as those developed to obtain certain target compounds
✓ New providers to supply the processes
✓ New retailers to sell the products
✓ New clients

Then, a large and diverse network of stakeholders can be built, generating sustainable wealth.

1.5 Relation between renewable carbon and green chemistry

To build a bridge between RC and green chemistry principles, an RC strategy should be closely related to principles to use renewable feedstock and design safer chemicals. However, other principles can also be considered according to the specificity and necessity to apply the RC strategy.

The 12 fundamental principles of green chemistry are as follows (ACS Green Chemistry Institute, 2021):

(I) Prevention: it is better to prevent waste than to treat or clean up waste after it has been created.

(II) Atomic economy: synthetic methods should be designed to maximize the incorporation of all materials used in the process into the final product.

(III) Less hazardous chemical syntheses: wherever practicable, synthetic methods should be designed to use and generate substances that possess little or no toxicity to human health and the environment.

(IV) Designing safer chemicals: chemical products should be designed to affect their desired function while minimizing their toxicity.

(V) Safer solvents and auxiliaries: the use of auxiliary substances (e.g., solvents, separation agents, etc.) should be made unnecessary wherever possible and innocuous when used.

(VI) Design for energy efficiency: energy requirements of chemical processes should be recognized for their environmental and economic impacts and should be minimized. If possible, synthetic methods should be conducted at ambient temperature and pressure.

(VII) Use of renewable feedstocks: a raw material or feedstock should be renewable rather than depleting whenever technically and economically practicable.

(VIII) Reduce derivatives: unnecessary derivatization (use of blocking groups, protection/deprotection, and temporary modification of physical/chemical processes) should be minimized or avoided, if possible, because such steps require additional reagents and can generate waste.

(IX) Catalysis: catalytic reagents (as selective as possible) are superior to stoichiometric reagents.

(X) Design for degradation: chemical products should be designed so that, at the end of their function, they break down into innocuous degradation products and do not persist in the environment.

(XI) Real-time analysis for pollution prevention: analytical methodologies need to be further developed to allow for real-time, in-process monitoring and control prior to the formation of hazardous substances.

(XII) Inherently safer chemistry for accident prevention: substances and the form of a substance used in a chemical process should be chosen to minimize the potential for chemical accidents, including releases, explosions, and fires.

The 12 principles are applicable, on a larger or smaller scale, to the production of chemicals and materials from RC materials. For example:

✓ Principles II, III, IV, VI, VII, VIII, IX, and X are directly useful for the development and production of chemicals.

✓ Principles I, V, XI, and XII are directly useful to renewable product usage.

Renewable feedstock should be understood as the main principle to be explored by RC strategies for petrochemicals substitution.

Then it is possible to observe that RC and green chemistry principles are complementary to each other to reach more sustainable chemistry.

1.6 Conclusions

The definition of RC is comprised of three components: biomass, carbon dioxide, and recyclable materials, which are largely produced and used by modern society.

Indeed, biomass, carbon dioxide, and recycled plastics are potential raw materials for several products, such as chemicals, materials, and pharmaceuticals, among others. The understanding of these three sources of RC and their physicochemical properties can start development of sustainable value chains to substitute for fossil raw materials (considering biomass and carbon dioxide) or to reuse those yet produced (considering the recyclable materials).

Moreover, the carbon atom is one of the most versatile constituents of organic molecules, acting as a skeleton for derived products, and is a perfect target for the application of green chemistry principles.

References

ACS Green Chemistry Institute, 2021. 12 Principles of Green Chemistry. https://www.acs.org/content/acs/en/greenchemistry/principles/12-principles-of-green-chemistry.html. access December 2021.

Charles, A.L., Sriroth, K., Huang, T.-C., 2005. Proximate composition, mineral contents, hydrogen cyanide and phytic acid of 5 cassava genotypes. Food Chem. 92, 615–620.

Chen, J., Zhang, L., Li, Q., Wang, M., Dong, Y., Yu, X., 2020. Comparative study on the evolution of polar compound composition of four common vegetable oils during different oxidation processes. LWT Food Sci. Technol. 129, 109538.

Clapp, C.E., Hayes, M.H.B., Senesi, N., Bloom, P.R., Jardine, P.M. (Eds.), 2001. Humic Substances and Chemical Contaminants. Soil Society of America, Madison.

Environmental Protection Agency, 2021. How do I Recycle? Common Recyclables. https://www.epa.gov/recycle/how-do-i-recycle-common-recyclables. access December 2021.

European Parliament, 2021. Circular Economy. https://www.europarl.europa.eu/thinktank/infographics/circulareconomy/public/index.html. access August 2021.

Falci, P.A., 2019. Repesando Práticas em Educação Ambiental: Proposta de uma Sequência Didática [Rethinking Environmental Educational Practices: Proposal for a Didactic Sequence]. M.Sc. thesis, Universidade Federal de Viçosa, Viçosa, Brazil, p. 48.

Faria, S., Petkowicz, C.L.O., De Morais, S.A.L., Terrones, M.G.H., De Resende, M.M., De França, F.P., Cardoso, V.L., 2011. Characterization of xanthan gum produced from sugar cane broth. Carbohydr. Polym. 86, 469–476.

Hepburn, C., Adlen, E., Beddington, J., Carter, E.A., Fuss, S., MacDowell, N., Minx, J.C., Smith, P., Williams, C.K., 2019. The technological and economic prospects for CO_2 utilization and removal. Nature 575, 87–97.

Intergovernmental Panel on Climate Change, 2021. Summary for policymakers. In: Climate Change 2021: The Physical Basis. IPCC 2021. Cambridge University Press, Cambridge.

International Energy Agency, 2021. CO_2 Emissions. https://www.iea.org/reports/global-energy-review-2021/co2-emissions. access December 2021.

Jenkinson, E.J., Adamns, D.E., Wild, A., 1991. Model estimates of CO_2 emissions from soil in response to global warming. Nature 351, 304–306.
Kweku, D.W., Bismark, O., Maxwell, A., Desmond, K.A., Danso, K.B., Oti-Mensah, E.A., Quachie, A.T., Adormaa, B.B., 2017. Greenhouse effect: greenhouse gases and their impact on global warming. J. Sci. Res. Rep. 17, 1–9.
Latham, K., 2021. The Way We Normally Recycle Plastics is a Downward Spiral of Waste and Degraded Materials, but there is Another Option—Turning Plastic Back into the Oil it was made from. https://www.bbc.com/future/article/20210510-how-to-recycle-any-plastic. access December 2021.
Liu, Q., Tarn, R., Lynch, D., Skjodt, N.M., 2007. Physicochemical properties of dry matter and starch from potatoes grown in Canada. Food Chem. 105, 897–907.
Mamma, D., Chistakopoulus, P., Koullas, D., Kekos, D., Macris, B.J., Koukios, E., 1995. An alternative approach to the bioconversion of sweet sorghum carbohydrates to ethanol. Biomass Bioenergy 8, 99–103.
Nalbant, Y., Colpan, C.O., 2020. An overview of hydrogen production from biogas. In: Uyar, T. (Ed.), Accelerating the Transition to a 100% Renewable Energy Era. Lecture Notes in Energy, vol. 74. Springer, Cham, pp. 355–373.
Organisation for Economic Co-operation and Development, 2018. Meeting Policy Challenges for a Sustainable Bioeconomy. https://www.oecd.org/publications/policy-challenges-facing-a-sustainable-bioeconomy-9789264292345-en.htm. access December 2021.
Poliakoff, M., Leitner, W., Streng, E.S., 2015. The twelve principles of CO_2 chemistry. Faraday Discuss. 183, 9–17.
Pubchem, 2022. Carbon Dioxide. https://pubchem.ncbi.nlm.nih.gov/compound/280 (accessed March 2022).
Renewable Carbon Initiative, 2021. The Renewable Carbon Initiative. https://renewable-carbon-initiative.com/. access December 2021.
Royal Society of Chemistry, 2021. Carbon. https://www.rsc.org/periodic-table/element/6/carbon. access December 2021.
Sandhu, K.S., Singh, N., Malhi, N.S., 2007. Some properties of corn grains and their flours I: physicochemical, functional and chapatti-making property of flours. Food Chem. 101, 938–946.
Sovová, K., Ferus, M., Matulková, I., Spanel, P., Dryahina, K., Dvorak, O., Civis, S., 2008. A study of thermal decomposition and combustion products of disposable polyethylene terephthalate (PET) plastic using high resolution fourier transform infrared spectroscopy, selected ion flow tube mass spectrometry and gas chromatography mass spectrometry. Mol. Phys. 106, 1205–1214.
Stevenson, F.J., 1994. Humus Chemistry: Genesis, Composition, Reaction, second ed. Wiley, New York.
Tatzber, M., Stemmer, M., Spiegel, H., Katzlberger, C., Hanernhauer, G., Gerzabek, M.H., 2008. Impact of different tillage practices on molecular characteristics of humic acids in a long-term field experiment—an application of three different spectroscopic methods. Sci. Total Environ. 406, 256–268.
The United Nations Food and Agriculture Organization, 2021. Crops and Livestock Products. http://www.fao.org/faostat/en/#data/QCL/visualize. access December 2021.
United Nations, 2021. Sustainable Development. https://sdgs.un.org/goals. access December 2021.
Vassilev, S.V., Baxter, D., Andersen, L.K., Vassileva, C.G., Morgan, T.J., 2012. An overview of the organic and inorganic phase composition of biomass. Fuel 94, 1–33.
Vaz Jr., S., 2016. Chapter 1 - The use of analytical chemistry to understand biomass. In: Vaz Jr., S. (Ed.), Analytical Techniques and Methods for Biomass. Springer Nature, Cham, pp. 3–5.

Vaz Jr., S., 2018. Analytical Chemistry Applied to Emerging Pollutants. Springer Nature, Cham.
Vaz Jr., S., 2020. Treatment of Agroindustrial Biomass Residues—A Sustainable Approach. Springer Nature, Cham.
von Braun, J., 2018. Bioeconomy: the global trend and its implications for sustainability and food security. Glob. Food Sec. 19, 81.
Wu, P., Huang, J., Zheng, Y., Yang, Y., Zhang, Y., He, F., Chen, H., Quan, G., Yan, J., Li, T., Gao, B., 2019. Environmental occurrences, fate, and impacts of microplastics. Ecotoxicol. Environ. Saf. 184, 109612.

CHAPTER 2

Organic synthesis for carbon derivatives

Organic chemistry, by means organic synthesis, has a very important role for the development and expansion of the oil-based economy. For example, the synthesis of polymeric materials—among many other products—is the basis of day-by-day modern society.

From the same point of view, organic synthesis should, again, have a highlighted role for the renewable carbon (RC) approach as discussed in this chapter.

2.1 Introduction to organic synthesis applied to renewable carbon products

Carbon-derived products are the basis of modern society. For example, the polymer industry is one of the most advanced sectors since the development of plastic polymer synthesis from oil. Moreover, it has expanded to several classes of molecules (e.g., acids, esters, alcohols, fatty acids, hydrocarbons, etc.) and applications, such as pharmaceuticals, agrochemicals, fuels, food, feed, and materials, among others.

Table 2.1 describes the bond enthalpies for several carbon-derived covalent bonds. It is important to point out that these covalent bonds are predominant in carbonaceous substances, which confer their chemical stability to them.

The bond enthalpy describes the amount of energy stored in a bond between atoms in a molecule. The higher the bond enthalpy, the more energy is needed to break the bond and, consequently, the stronger the bond. It is closely related to the mean bond energy definition of International Union of Pure and Applied Chemistry (2014), i.e., the average value of the gas-phase bond dissociation energies.

Renewable Carbon
https://doi.org/10.1016/B978-0-323-99735-5.00014-1

Table 2.1 Bond enthalpies for covalent bonds in carbon-derived organic compounds.

Covalent bond	Enthalpy ($kJ\,mol^{-1}$)	Found in
C—C	345.6	General chemical species
C=C	610	General chemical species
C≡C	835.1	General chemical species
C—N	304.6	General chemical species
C=N	615	General chemical species
C≡N	889.5	General chemical species
C—F	485	CF_4
C—Si	301	$(CH_3)_4Si$
C—Cl	339	General chemical species
C—Cl	327.2	CCl_4
C—I	218	General chemical species
C—I	213	CH_3I
C—Br	285	General chemical species
C—H	413	General chemical species
C—H	415.5	CH_4
C—O	357.7	General chemical species
C=O	803	CO_2
C=O	695	HCHO
C=O	736	Aldehydes
C=O	749	Ketones
C—O	335.6	CH_3OH

Adapted from the Royal Society of Chemistry, 2021. Carbon. https://www.rsc.org/periodic-table/element/6/carbon, Accessed December 2021. Reprinted with permission from the Royal Society of Chemistry.

As observed by Kühlborn et al. (2020), along with the utilization of biomass for fuel and energy production, the search for new sources of chemical feedstocks is a growing field in current chemical research. Numerous recent publications deal with the production of low molecular weight compounds from biomass and the valorization of these renewable feedstocks for synthetic routes, although commodities such as adipic acid, acrylic acid, and butanol currently still dominate the field. Furthermore, since organic synthesis is about constructing carbon skeletons, the origin and fate of carbon atoms significantly contributes to the sustainability of certain organic products. Thus, the renewability of raw materials, i.e., precursors, is a major aspect of green chemistry attended by RC sources.

Then the use of biomass to promote green chemistry, by means of synthetic routes, will generate new products, technological processes, and value chains to present alternatives to fossil raw materials (Vaz Jr., 2018). Moreover, it can be applied to CO_2-derived products. Nevertheless, it is very limited or impracticable for recyclable materials due their physical properties (e.g., physical state) and impurities or contaminants added during the processing step.

Generally, when choosing a certain organic reaction for RC raw material conversion into a chemical or a new material, one should consider:

- The chemical constitution of the raw material and the presence of contaminants
- The purity to be obtained for the final product as well as its yield (in mol or mass)
- The laboratory facility (e.g., analytical instrumentation, glassware, reactors, reactants, etc.)
- The necessity of catalysts—it can make the reaction very expensive
- And more recently, which principles of green chemistry can be applied (e.g., atomic economy, energy economy, design of environmentally friendly molecules, etc.).

Examples of organic reactions to be applied to obtain products from biomass are depicted in Fig. 2.1.

The main advantage of biomass materials is that they will generate oxygen-derived products. However, the high oxygen content will need a reductive step when the goal is to obtain hydrocarbons and related compounds.

On the other hand, examples of reactions to be applied to obtain products from carbon dioxide are (Yao et al., 2020):

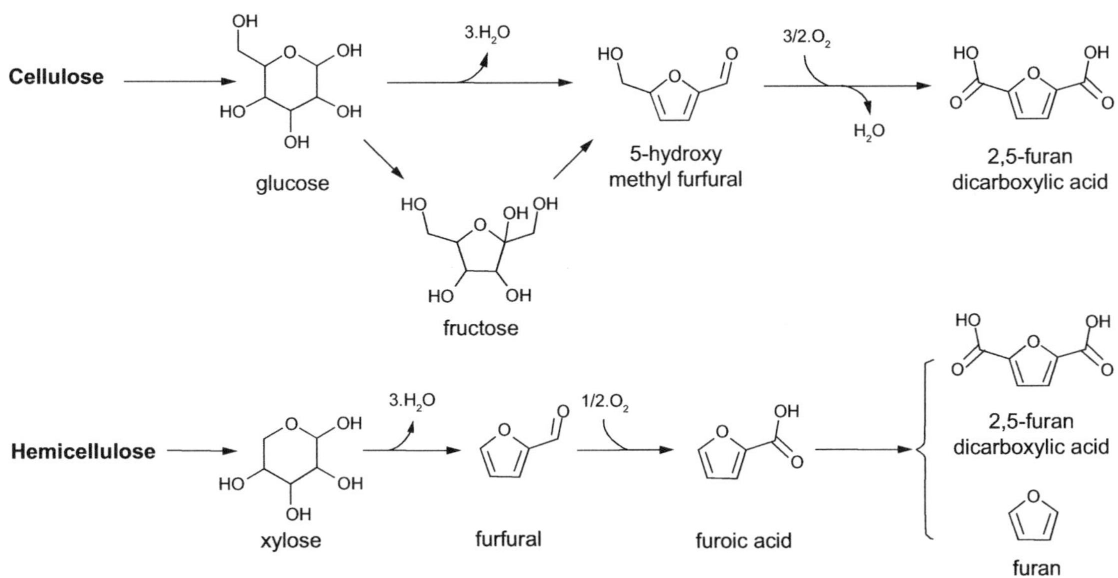

Fig. 2.1 The production of 2,5-furandicarboxilic acid (FDCA) from cellulose and hemicellulose. *(Credit: Iglesias, J., Martínez-Salazar, I., Maireles-Torres, P., Martin Alonso, D., Mariscal, R., 2020. Advances in catalytic routes for the production of carboxylic acids from biomass: a step forward for sustainable polymers. Chem. Soc. Rev. 16, 5704–5771. Reproduced with permission from the Royal Society of Chemistry.)*

Hydrogenation (Scheme 2.1):

$$CO_2 + 3H_2 \rightleftarrows -(CH2)- + 2H_2O\ (\Delta_f H^{\circ}_{298} = -\ 125\ kJ\ mol^{-1})$$

Scheme 2.1

The reverse water gas shift reaction (Scheme 2.2):

$$CO_2 + H_2 \rightleftarrows CO + H_2O\ (\Delta_f H^{\circ}_{298} = +41\ kJ\ mol^{-1})$$

Scheme 2.2

The Fisher-Tropsch synthesis (Scheme 2.3):

$$CO + 2H_2 \rightleftarrows -(CH_2)- + H_2O\ (\Delta_f H^{\circ}_{298} = -\ 166\ kJ\ mol^{-1})$$

Scheme 2.3

We can combine different sources of RC. For example, a combination of carbon dioxide and lignin from lignocellulosic biomass, among other reactants and catalysts, can produce ethanol (Fig. 2.2).

Furthermore, CO_2-based polymers are an interesting use of this gas as a raw material (Fig. 2.3).

Carbon dioxide emerges as a raw material for several chemical products currently manufactured from fossil materials, e.g., crude oil and naphtha. Of course, it is an opportunity to be best explored by academia and industries with scientific and economic criteria, but it can promote a net-zero emission for industries with economic, social, and environmental gains. Regarding net-zero emission, chemical industry is focused on energy efficiency and renewable power now, but it will likely need to examine carbon capture

$$\text{Lignin-}(OCH_3)_n + CO_2 + H_2 \xrightarrow[\geq 140\ ^{\circ}C,\ \text{in DMI}]{\text{Ru-Co-LiI-triphos}} CH_3\text{-}CH_2\text{-}OH$$

Fig. 2.2 Synthesis of ethanol from lignin, CO_2, and H_2. *(Credit:* Zhang, J., Qian, Q., Wang, Y., Asare-Bediako, B., Yan, J., Han, B., 2019. Synthesis of ethanol from aryl methyl ether/lignin, CO2 and H2. Chemical Science 10, 10640–10646. https://doi.org/10.1039/C9SC03386F. *Reprinted with permission from the Royal Society of Chemistry.)*

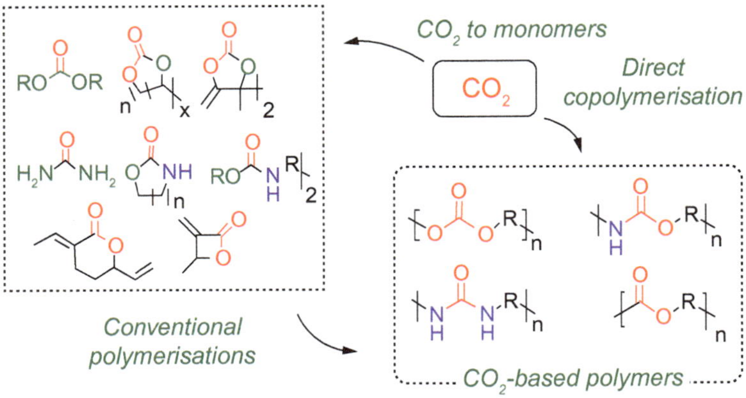

Fig. 2.3 CO_2-based monomers and polymers. *(Credit: Grignard, B., Gennen, S., Jérome, C., Kleij, A.W., Detrembleur, C., 2019. Advances in the use of CO2 as a renewable feedstock for the synthesis of polymers. Chem. Soc. Rev. 48, 4466–4514. Reproduced with permission from the Royal Chemistry Society.)*

at some point (Bettenhausen, 2021), using carbon capture and use (CCU) technologies involving organic and inorganic synthesis.

Regarding recyclable materials, generally, they need a physical or thermochemical process for reuse to obtain new products; reactions needed for this raw material are in detail in Chapter 10.

2.2 Organic processes for the synthesis of renewable carbon products

A synthetic strategy is depicted in Fig. 2.4 for RC raw materials. Initially, the RC source must undergo a complete chemical analysis and characterization, which aims to determine its chemical constitution in addition to some physicochemical properties of interest. Then, there is the treatment of the starting material, when necessary, which will allow the separation of the precursor molecule of interest; if it does not have adequate purity, a purification step is carried out. With the obtaining of the precursor molecule, an organic synthesis step is taken, in which the search for the best catalysts may be involved, that is, the catalytic screening of various catalysts, such as heterogeneous and homogeneous inorganic catalysts, organometallic catalysts and enzymes, and the appropriate approach for design of the synthesis routes.

After the synthesis of the target product, it must be properly identified as to its chemical structure and purity by means of analytical techniques as ^{13}C or ^{1}H NMR, MS, and FTIR. After chemical identification, study of the

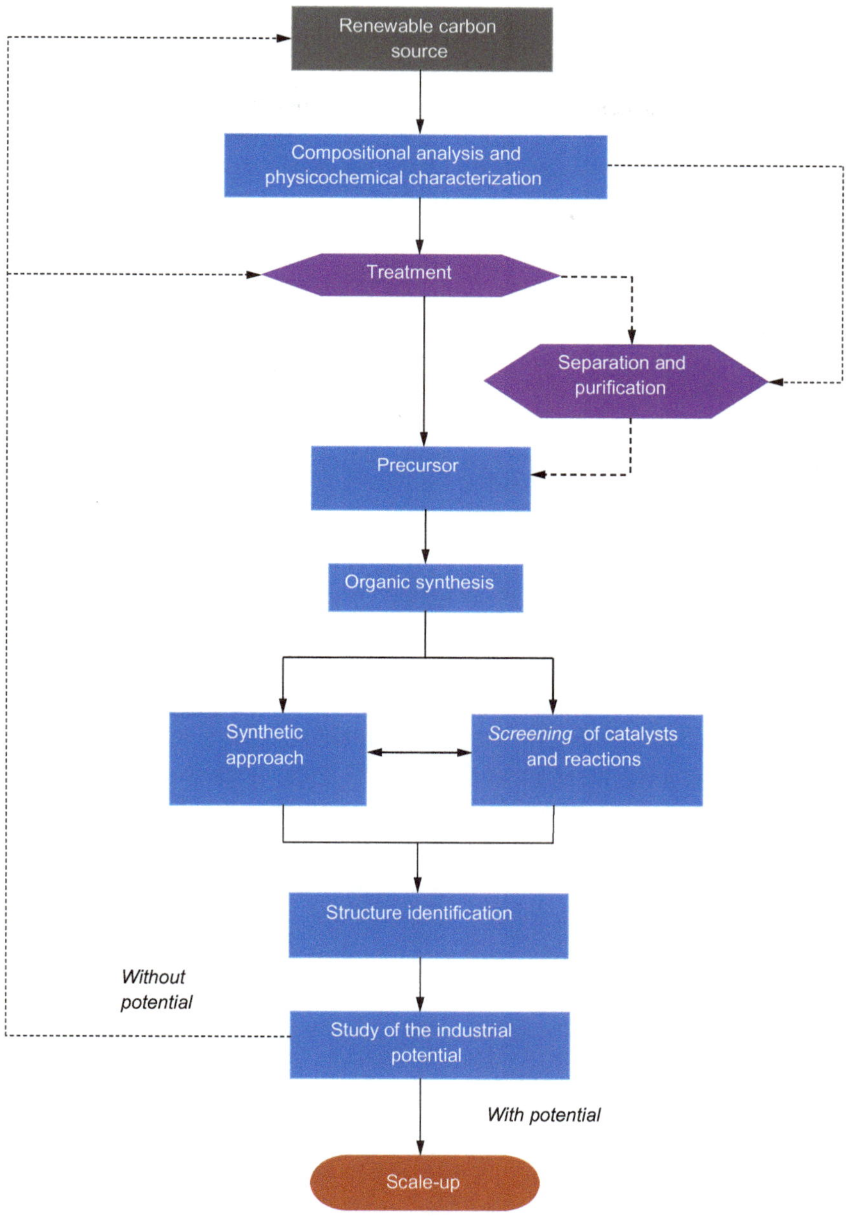

Fig. 2.4 Flowchart of an organic process applied to renewable carbon source. *(Credit: author.)*

Table 2.2 Products obtained from RC raw materials.

RC raw material	Renewable products
Carbon dioxide	Cyclic carbonates, organic carbamates, salicylic acid, formic acid, polycarbonate-etherols, poly(propylene) carbonate, urea, methanol, DME, aldehydes, alcohols
Lignin from biomass	Syngas products (methanol, DME, ethanol, C1–C7 gases, etc.), hydrocarbons (BTEX, etc.), phenols and polyphenols (substituted phenols, cresols, catechols, etc.), oxidized products (vanillin, DMSO, aromatic acids, aliphatic acids, quinones, etc.), macromolecules (carbon fiber, substituted lignins, composites, pharmaceuticals, etc.)
Sugars from biomass	Levulinic acid, FDCA, monoethylene glycol, methyl vinyl glycolate

DME, 1,2-dimethoxyethane; *BTEX*, benzene, toluene, ethylbenzene and xylenes; *DMSO*, dimethyl sulfoxide; *FDCA*, 2,5-furandicarbolylic acid.
Source: Adapted from International Energy Agency, 2020. Bio-Based Chemicals—A 2020 Update. IEA Bioenergy, 79 p.

target product's potential and its synthesis route is carried out; when it presents industrial potential, the next stage is the scale-up with a view to industrial production. If product and route are not economically viable, the search for a new precursor molecule, a new target product, or both can be restarted.

It is worth commenting on the development and use of catalysts for these processes (discussed in detail in Chapter 11), given their importance for improving yield and selectivity, by considering enantioselectivity, regioselectivity, and stereoselectivity aspects.

Table 2.2 describes several products from RC raw materials obtained by means of organic synthesis.

Some of the types of reactions to be applied for RC raw material conversion are depicted in Table 2.3.

A detailed description of each type of reaction in Table 2.3 can be obtained in books of synthetic organic chemistry, such as *March's Advanced Organic Chemistry Reactions, Mechanisms, and Structures.*[a]

Considering that RC raw materials are highly heterogeneous matrices, they can demand a combination of reactions to reach a target product. Indeed, an organic reaction can be combined to reach the best potential of the feedstocks, as observed by Bozell and Petersen (2010) and shown in Fig. 2.5 for succinic acid.

[a] https://www.wiley.com/en-us/March%27s+Advanced+Organic+Chemistry%3A+Reactions%2C+Mechanisms%2C+and+Structure%2C+8th+Edition-p-9781119371809.

Table 2.3 Examples of organic reactions for RC raw materials conversion.

RC raw material	Organic reactions
Carbon dioxide	Electrolysis Hydrogenation Rearrangement
Lignin from biomass	Molecular cracking in the presence of catalysts (e.g., metallic catalysts, enzymes) in oxidative or reductive medium
Sugars from biomass	Aldolic condensation Oxidation Reduction Rearrangement
Other molecules	Acetalization Chlorination Cyclization Dehydrogenation Dehydration Etherification Esterification Tautomerism Hydration

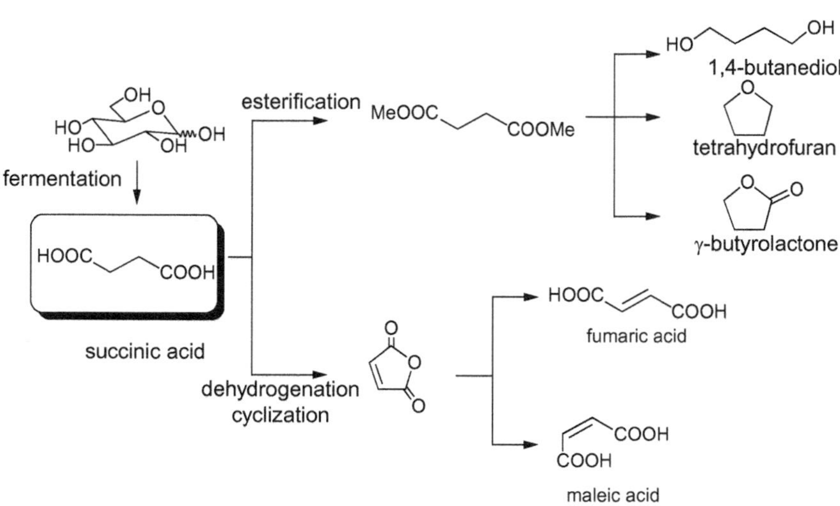

Fig. 2.5 Succinic acid as a platform chemical. *(Source: Bozell, J.J., Petersen, G.R., 2010. Technology development for the production of biobased products from biorefinery carbohydrates—the US Department of Energy's "Top 10" revisited. Green Chem. 12, 539–554. Reproduced with permission from the Royal Society of Chemistry.)*

Succinic acid has a role as a nutraceutical, a radiation protective agent, an antiulcer drug, a micronutrient, and a fundamental metabolite. Moreover, it is a flavoring agent and can be used to synthesize several products, such as furans, acids, and alcohols.

The reaction control for synthetic or conversion routes can take place as:

- Thermodynamic control: obtaining the most stable products, which leads to lower values of Gibbs free energy (ΔG) for the reaction
- Kinetic control: obtaining products with lower activation energy (E_{at}) and higher reaction speed

The thermodynamics applied to conversion routes considers the free energy of activation ($\Delta G^{\#}$), and its decrease favors the formation of products (Eq. 2.1):

$$\Delta G^{\#}\left(\text{kJ}\,\text{mol}^{-1}\right) = \Delta H^{\#} - T\Delta S^{\#} \tag{2.1}$$

where $\Delta H^{\#}$ is the activation enthalpy (kJ mol^{-1}), T is the system temperature (K), and $\Delta S^{\#}$ the activation entropy (J K^{-1} mol^{-1}).

For the reaction kinetics, in a hypothetical reaction where reactants A and B lead to the formation of product C in the presence of a catalyst P, the reaction rate can be expressed as follows (Eq. 2.2):

$$\nu\left(\text{s}^{-1}\,\text{mol}\,\text{L}^{-1}\right) = k[A][B][P] \tag{2.2}$$

where k is the rate constant (s^{-1}, for first-order reactions; or dm^{3} mol^{-1} s^{-1}, for second-order reactions), $[A]$ and $[B]$ are the reactant concentrations (mol L^{-1}), and $[P]$ is the catalyst concentration (mol L^{-1}).

Another simplified way to express the kinetics of a reaction is (Eq. 2.3):

$$\nu\left(\text{mol}\,\text{L}^{-1}\,\text{s}^{-1}\right) = \text{d}[C]/\text{dt} \tag{2.3}$$

where d$[C]$ is the concentration of the product formed (mol L^{-1}) from its first derivative, and dt is the reaction time (s) also from its first derivative.

However, the reaction rate can be directly proportional to the catalyst concentration (Eq. 2.4):

$$\nu\,\alpha\,[P] \tag{2.4}$$

Therefore, the reaction rate can be inversely proportional to the activation energy of the reaction (Eq. 2.5):

$$\nu\,\alpha\,1/E_{at} \tag{2.5}$$

2.3 Industrial carbon-based organic chemicals: A case study

Glycerol is a by-product obtained from biodiesel production. Its conversion into lactic acid and methyl lactate by means of organic synthesis is depicted in

Fig. 2.6. We can observe several reactions involved as dehydrogenation, dehydration, tautomerism, acetalization, and hydration.

Lactic acid is a liquid used as a flavoring agent, plant growth regulator, and drug addictive. Its ester methyl lactate is also a liquid used as solvent for cellulose derivatives, as cellulose acetate, and nitrocellulose. Currently, it is produced via fermentation by different microorganisms which can use different carbohydrate sources, such as the lactic acid bacteria *Lactobacillus* spp. (Abedi and Hashemi, 2020). When comparing organic synthetic routes against biochemical synthetic routes, its advantages are:

✓ Production of high-purity grade product
✓ High reaction rate

In general, organic syntheses will provide the best purity and yield a higher reaction rate when compared against biochemical syntheses. However, their application will depend on economic aspects.

2.4 Industrial carbon-based organic materials: A case study

Borregaard (2021) is a European company based on wood components and is a pioneer in lignin chemical businesses in the United States and Sweden for more than 90 years.

According to the company, their lignin-based macromolecules—lignosulfonates as seen in Fig. 2.7—are renewable, wood-based alternatives to fossil-based chemicals and polymers used in a broad range of industries. These products are tailor-made to obtain optimal performance in their application as environmentally friendly products used as binders, dispersing agents, crystal growth modifiers, emulsion stabilizers, and complexing agents (Fig. 2.8). These multifunctional properties are valuable in a vast range of applications.

The raw material is wood without competing with food production. Documentation of the sustainability of these products and processes is spearheaded by a Life Cycle Analysis (LCA)[b] approach providing Environmental Product Declarations (EPDs),[c] as discussed in Part 3.

[b] https://www.iso.org/standard/37456.html.
[c] https://www.iso.org/obp/ui#iso:std:iso:14025:ed-1:v1:en.

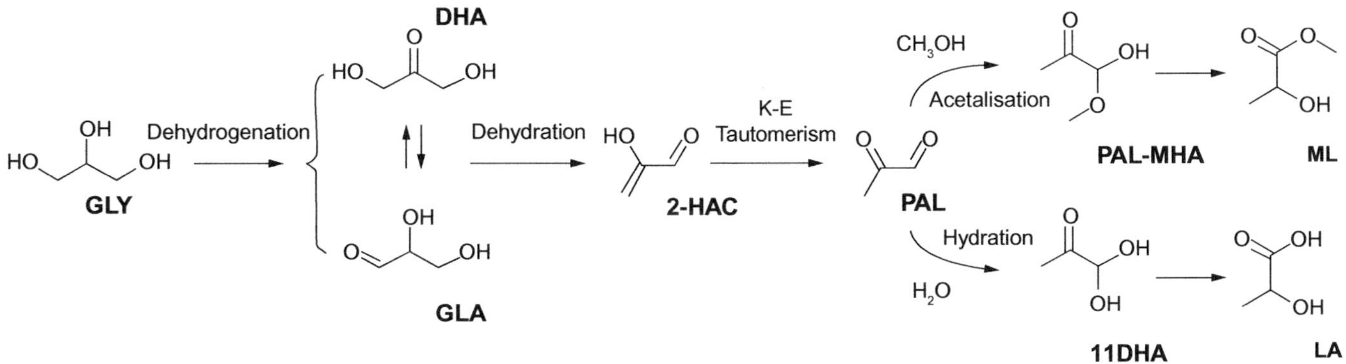

Fig. 2.6 Transformation of glycerol into lactic acid and methyl lactate. *GLY*, glycerol; *DHA*, dihydroxyacetone; *GLA*, glyceraldehyde; *2-HAC*, 2-hydroxy acrolein; *PAL*, pyruvaldehyde; *PAL-MHA*, pyruvaldehyde methyl hemiacetal; *ML*, methyl lactate; *1,1-DHA*, 1,1-dihydroxy acetone; *LA*, lactic acid. *(Credit: Iglesias, J., Martínez-Salazar, I., Maireles-Torres, P., Martin Alonso, D., Mariscal, R., 2020. Advances in catalytic routes for the production of carboxylic acids from biomass: a step forward for sustainable polymers. Chem. Soc. Rev. 16, 5704–5771. Reproduced with permission from the Royal Society of Chemistry.)*

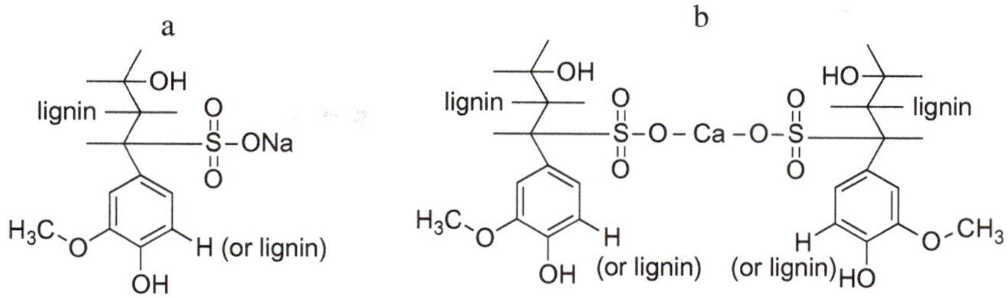

Fig. 2.7 Lignosulfonate structures (A and B). *(Source: Lugovitskaya, T.N., Naboychenko, S.S., 2020. Lignosulfonates as charge carriers and precursors for the synthesis of nanoparticles. Colloids Surf. A 602, 125127. Reproduced with permission from Elsevier.)*

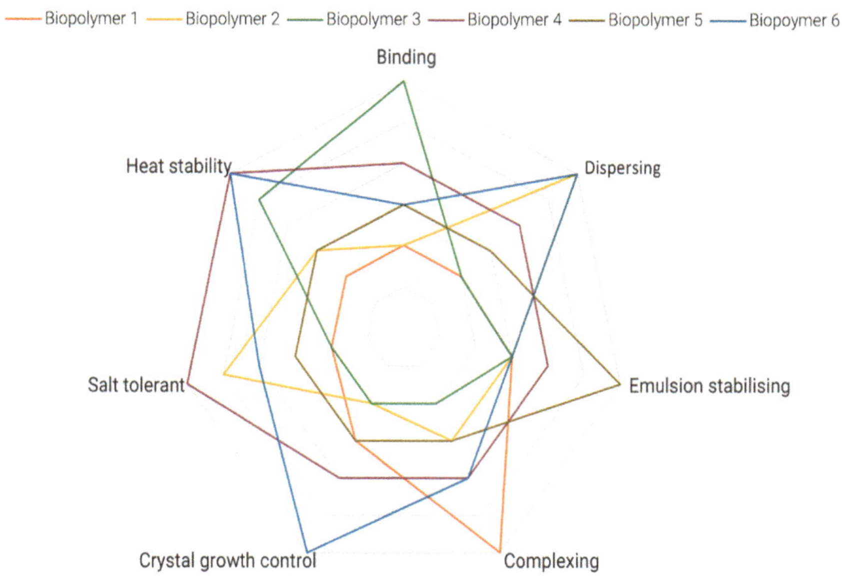

Fig. 2.8 Application of biopolymers derived from lignin. *(Courtesy of Borregaard.)*

2.5 Conclusions

Organic synthesis presents a huge contribution to reach the RC concept. Its reactions are paramount to obtain renewable products, i.e., chemicals and materials, because a majority of these products are organic molecules that need a conversion step applied to feedstock materials.

Biomass and carbon dioxide are the main sources of RC for organic synthesis; recyclable materials generally need a physical or thermochemical process for their reuse, making new polymerization reactions unfeasible.

The heterogeneity of RC raw materials, especially biomass, point to a necessity for a combination of organic reactions to reach the best exploitation of these feedstocks.

References

Abedi, E., Hashemi, S.M.B., 2020. Lactic acid production—producing microorganisms and substrates sources-state of art. Heliyon 6, e04974.

Bettenhausen, C., 2021. Climate change—chemical firms aim for more ambitious climate goals. Chem. Eng. News 99, 33.

Borregaard, 2021. Lignin Biopolymers. https://www.borregaard.com/product-areas/lignin-biopolymers/?utm_source=google%20&utm_medium=cpc&utm_term=lignin%20biopolymers&utm_campaign=lignin%20biopolymers&utm_term=lignins&utm_campaign=Product+Area+%7C+Lignin+Biopolymers+%7C+Global&utm_source=adwords&utm_medium=ppc&hsa_acc=3642184401&hsa_cam=13555587785&hsa_grp=126240300920&hsa_ad=528203487243&hsa_src=g&hsa_tgt=kwd-328562444388&hsa_kw=lignins&hsa_mt=p&hsa_net=adwords&hsa_ver=3&gclid=CjwKCAjw95yJBhAgEiwAmRrutLbJnlGrT3OFab2T8SpnI6PoZdG_0K9Ys0TmeUC2tkOQO0LctNeqIRoCRtkQAvD_BwE. accessed December 2021.

Bozell, J.J., Petersen, G.R., 2010. Technology development for the production of biobased products from biorefinery carbohydrates—the US Department of Energy's "Top 10" revisited. Green Chem. 12, 539–554.

International Union of Pure and Applied Chemistry, 2014. Bond Energy. https://goldbook.iupac.org/terms/view/B00701. accessed December 2021.

Kühlborn, J., Groβ, J., Opatz, T., 2020. Making natural products from renewable feedstocks: back to the roots? Nat. Prod. Rep. 37, 380–424.

Vaz Jr., S. (Ed.), 2018. Biomass and Green Chemistry—Building a Renewable Pathway. Springer Nature, Cham.

Yao, B., Xiao, T., Makgae, O.A., Jie, X., Gonzalez-Cortes, S., Guan, S., Kirkland, A.I., Dilworth, J.R., Al-Megren, H.A., Alshihri, S.M., Dobson, P.J., Owen, G.P., Thomas, J.M., Edwards, P.P., 2020. Transforming carbon dioxide into jet fuel using an organic combustion-synthesized Fe-Mn-K catalyst. Nat. Commun. 11, 6395.

CHAPTER 3

Inorganic and thermochemical syntheses for carbon derivatives

Inorganic and thermochemical syntheses are subjects of interest, development, and application for a large number of industrial sectors. However, their potential for production of carbonaceous materials is not well explored, being discussed in this chapter.

It is expected that the partnership between both approaches could promote the expansion of the carbon derivatives.

3.1 Introduction to inorganic and thermochemical syntheses applied to renewable carbon products

As well as organic synthesis, inorganic synthesis has a purpose for the conversion of a certain raw material (precursor) into a final product (target compound). Here, we are considering renewable carbon (RC) sources as the raw material.

Although organic synthesis can be applied to biomass and carbon dioxide, inorganic synthesis is restricted to carbon dioxide as a RC raw material. This is because CO_2 is an inorganic molecule generating inorganic chemicals and materials as final products, except for its polymerization to polycarbonates as polyether (Langanke et al., 2015). However, the thermochemical approach, as a complementation of the first one or as a sole approach, routes of synthesis for some interesting materials from biomass and renewable materials (e.g., industrial polymers).

Examples of products obtained by inorganic and/or thermochemical synthesis include:

- Bicarbonate, obtained by inorganic synthesis
- Carbonate and nanocarbonate, obtained by inorganic synthesis
- Graphene, obtained by thermochemical synthesis
- Graphene oxide, obtained by an association of inorganic synthesis with thermochemical synthesis
- Microporous carbon, obtained by thermochemical synthesis (carbonization)
- Mesoporous carbon, obtained by thermochemical synthesis (carbonization)

https://doi.org/10.1016/B978-0-323-99735-5.00015-3

- Hierarchical carbon, obtained by association of inorganic synthesis (ion exchange) with thermochemical treatment
- Activated carbon, obtained by thermochemical synthesis (carbonization)
- Carbon nanotubes, obtained by thermochemical synthesis (pyrolysis)
- Fullerene, obtained by the thermal treatment of graphite with a high-power laser

Fig. 3.1 depicts the structure of fullerene as an example of carbonaceous materials obtained from RC.

Fullerene can be more conveniently produced by using an electric arc between graphite electrodes in an inert gas, followed by solubilization in benzene, then crystallization to form a solid (yellow crystals). Its structure can be used to entrap some chemical and biological species, such as toxic metals and pathogens (Azizi-Lalabadi et al., 2020; Baby et al., 2019).

Regarding inorganic synthetic routes, we can highlight some reactions:

- Base-catalyzed reaction using a base as catalyst
- Bosch process, the reaction between CO and steam over a hot heterogeneous catalyst
- Ion-exchange reaction, with exchange cations, such as H^+ and Na^+, for cations in the surrounding medium
- Electrochemical synthesis
- Thermal decomposition of $CaCO_3$

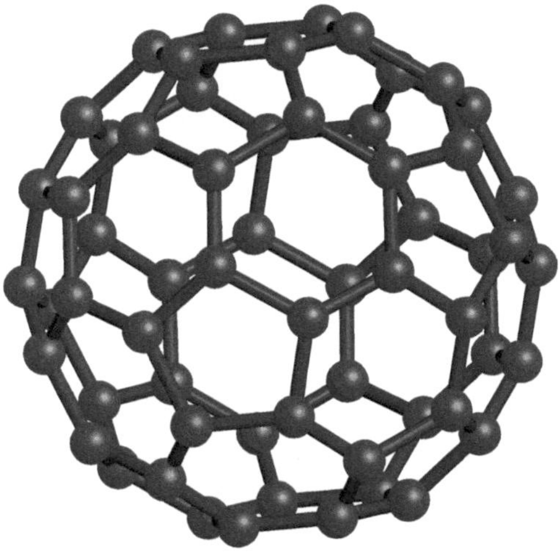

Fig. 3.1 The carbon allotrope fullerene (or buckminsterfullerene) that contains a cluster of 60 carbon atoms linked in a highly symmetrical polyhedral structure. *(Credit: Shutterstock, https://www.shutterstock.com/pt/image-vector/fullerene-c60-ball-stick-model-3d-1921625939.)*

Regarding thermochemical synthesis, the main synthetic routes involve:

- Combustion: the raw material is exposed to moderate to high temperature (up to 1600°C) in the presence of oxygen with CO_2 generation; incomplete combustion of carbonaceous materials can generate polyaromatic hydrocarbons (PAHs) molecules that are carcinogenic (Boffetta et al., 1997; Petit et al., 2019).
- Gasification (Fig. 3.2): the raw material is exposed to high temperature and pressure in the absence of oxygen or in very poor oxygen content with the generation of CO and H_2 (syngas).
- Pyrolysis: the raw material is exposed to high temperature (>600°C) in absence of oxygen or in very poor oxygen content at a short time of residence (seconds) with the generation of reduced products as hydrocarbons (C_nH_n). The slow pyrolyzation at a high time of residence is known as *carbonization*.
- Torrefaction (or mild pyrolysis): the raw material is exposed to moderate temperature (250–280°C) in absence of oxygen or in very poor oxygen content. This process maintains the burn properties of the treated raw material, e.g., biomass.

Thermochemical technologies are very common for petrochemical and biomass industries. In some cases, it can be applied to degrade plastic vessels (recyclable materials) contaminated by chemical residues, such as pesticides, as a treatment technology. Nevertheless, this set of technologies are high-energy dependent with increasing costs and without a direct contribution to green chemistry, i.e., the principle of energy economy.

Table 3.1 presents a perspective of application for the most advanced products obtained from RC by inorganic and/or thermochemical synthesis.

Fig. 3.3 depicts the structure of carbon nanotubes.

Nanotechnology and their synthetic approaches (e.g., solvents, spray dryer, precipitation, etc.) can be explored by inorganic and thermochemical routes to achieve improved properties, such as:

- Active sites for intermolecular interactions
- Pore volume for adsorption and absorption
- Surface area for adsorption

3.2 Inorganic and thermal processes for the synthesis of renewable carbon products

As previously stated, carbon dioxide is the best example for inorganic synthesis to obtain renewable products from an RC source.

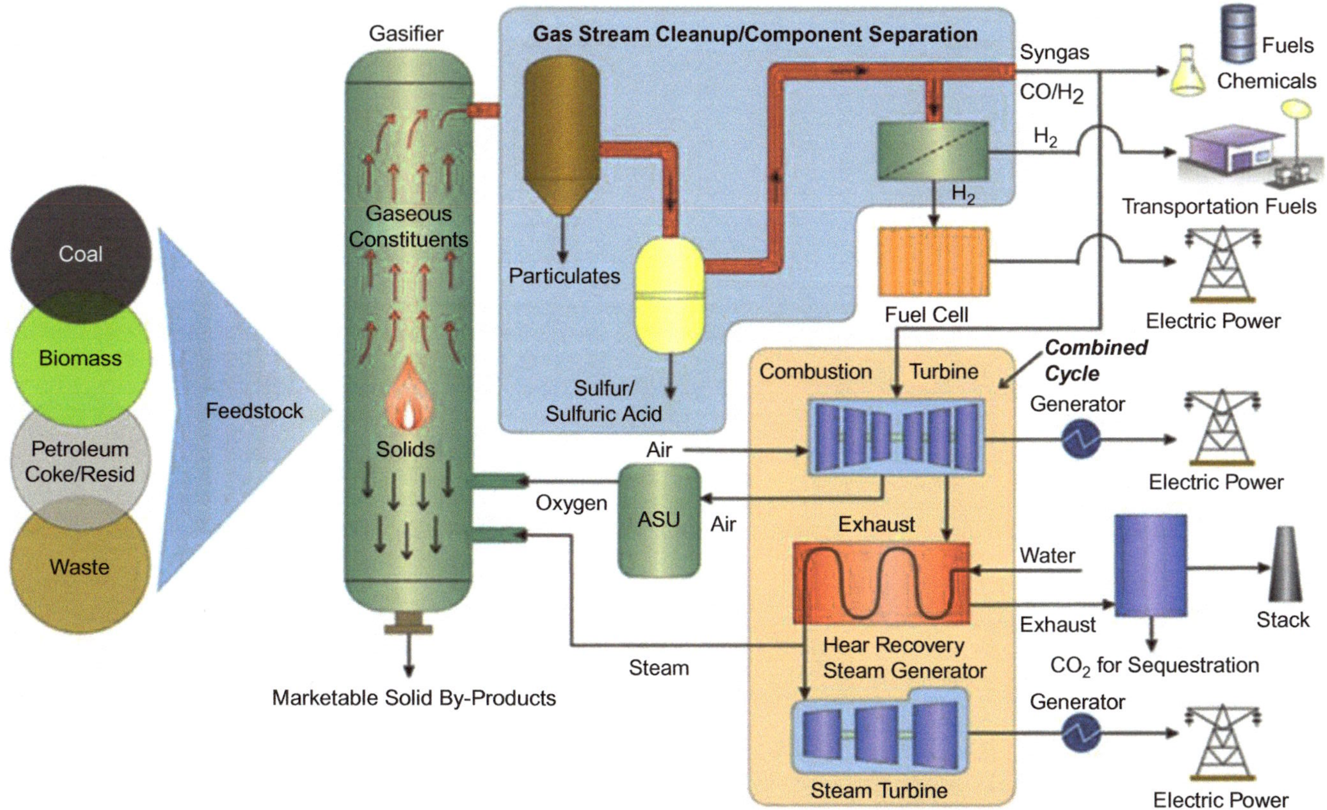

Fig. 3.2 Illustration of the gasification process for RC sources but not limited to just that. *ASU*, air separation unity. *(Reproduced with permission from the U.S. Department of Energy.)*

Table 3.1 Advanced products obtained from RC by inorganic or thermochemical synthesis.

Carbon-based renewable product	Synthetic route	Main characteristic(s)	Application(s)	Reference
Bicarbonate	Inorganic	Powder	Chemical reactant	Bonaventura et al. (2017)
Carbonate	Inorganic	Powder	Chemical reactant	Bonaventura et al. (2018)
Nanocarbonate	Inorganic	Powder with porous vacancies (relatively small)	Drug delivery carrier	Palmkvist et al. (2017)
Activated carbon	Thermochemical (pyrolysis)	Powder with available pores for absorption	Filter element, separations, environmental include for CO_2 absorption	Idrees et al. (2018)
Carbon nanotubes	Thermochemical (microwave-assisted pyrolysis)	Cylindrical nanostructure with high thermal and electrical conductivity and low thermal expansion coefficient	Absorption, filtration, separation, environmental, catalysis	Omoriyekomwan et al. (2019)
Graphene	Thermochemical (vapor deposition of H_2 and CH_4)	3D powder sheets	Advanced material for energy storage	Ji et al. (2021)

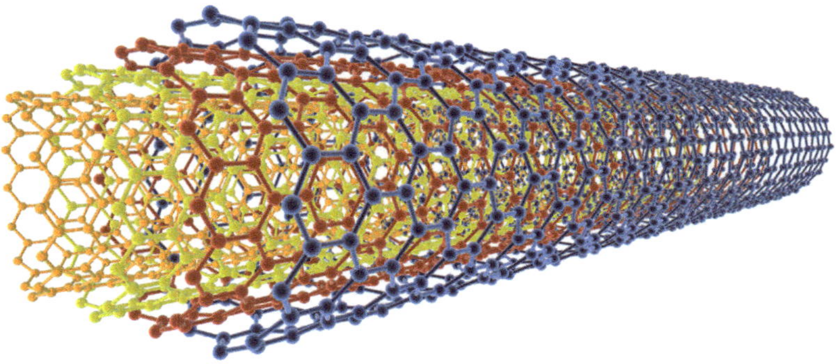

Fig. 3.3 Computational model of a multiwalled carbon nanotubes. *(Credit: Shutterstock, https://www.shutterstock.com/pt/image-illustration/multiwalled-carbon-nanotubes-mwnts-russian-doll-474371770.)*

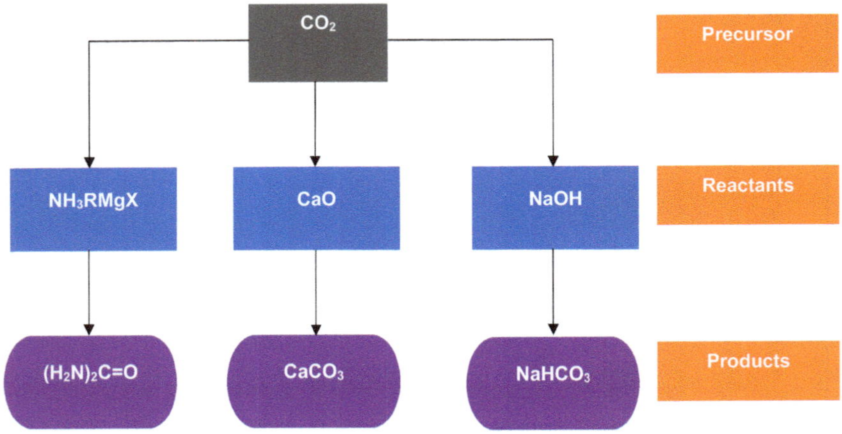

Fig. 3.4 An inorganic product tree from CO_2: urea *(left)*, sodium carbonate *(middle)*, and sodium bicarbonate *(right)*. *(Credit: author.)*

Fig. 3.4 depicts some examples of products obtained from CO_2. Their processes can be considered for carbon capture and use (CCU) because the CO_2 can be chemically captured by its conversion to a certain product, which could be used as an input or final product.

The usages of the products presented in Fig. 3.4 are:

- Urea ($(H_2N)_2C{=}O$) is used in agriculture and livestock.
- Sodium carbonate ($CaCO_3$) is used in agriculture, chemical industry, and pharmaceutical industry, among others.
- Sodium bicarbonate ($NaHCO_3$) is used in chemical laboratories as a reactant or for pH control, and in chemical and pharmaceutical industries, among others.

These chemicals have a characteristic of net-zero carbon—an overall balance between greenhouse gas (GHG) emissions produced and GHG emissions taken out of the atmosphere by a certain process—if we consider combustion of fossil fuels as the source of CO_2. This environmental approach is discussed in detail in Part 3 for sustainability of products and processes.

Regarding the previously discussed thermal (or thermochemical) processes, with the exception of combustion, the other thermochemical processes of RC conversion are all carried out in the absence or deficiency of oxygen, bearing in mind that its total absence is very difficult to obtain in practice. Thus, the difference between the processes lies in operating temperature and pressure, residence time, operating regime, particle size, and catalyst usage.

Fast pyrolysis is so-called because of its short residence time, which is the time it takes for reactants (or raw materials) to enter, react, form products, and leave the reactor. Fast pyrolysis is also so-called to differentiate from slow pyrolysis, more commonly known as carbonization. The residence time in rapid pyrolysis is of the order of seconds, that is, the RC takes from 2 to 3 s to transform into products. In carbonization, the same cycle can last a few weeks. Such difference is due to the mechanisms of heat and mass transfer, which in fast pyrolysis is instantaneous, and in carbonization, it takes a long time. The main influence on the process is the particle size, as the particles subjected to rapid pyrolysis must be at most 1 mm, and in the case of carbonization, the particles have of 1 m in length by several inches in diameter. These are fundamental differences between rapid pyrolysis and carbonization, but beyond that, the products are also quite different in the two processes. Carbonization mainly produces charcoal, and rapid pyrolysis has bio-oil as its main product. However, solid, liquid, and gaseous products are always formed in these processes, regardless of the technology applied. Often, some of these products are not recovered or used and are simply discarded. This is the case of charcoal produced in masonry ovens, which wastes the liquid and gaseous products of carbonization in the form of atmospheric emissions. However, the emissions and pollutant generation can be reduced by means of carbonization based on resistive heating (von Broembsen, 2007) to produce high-grade activated carbon from biomass waste (e.g., hardwood, bamboo, coconut, etc.); the obtained activated carbon can be used as purification technology for water, air, food, beverages, pharmaceuticals, and industrial pollutants (organic and inorganic species).

Roasting, also called mild pyrolysis, is a thermal treatment that takes place between 250°C and 280°C; although it has been known for a long time, there is recent interest in processes that preserve most of the RC for later burning and require high calorific value. For example, roasting produces roasted biomass, which is an intermediate solid fuel between fresh biomass and charcoal. Roasted biomass goes beyond biomass drying and initiates the thermal degradation process, interrupting itself in the endothermic phase of pyrolysis. This process produces a high-quality biofuel, with high density and better burning properties. Roasting biomass could become a popular process in the future.

Gasification is already a widely used process with many technologies associated with it. RC gasification still has a lot of development and innovation until it reaches the market in a comprehensive way; however, fossil fuel gasification is widely used in industry. Coal, heavy oil fractions, and even natural gas are all currently gasified; gasify means to transform the initial fuel into a synthesis gas ($CO + H_2$, majority). Gasification has many applications, such as thermal purposes, electrical generation, synthesis of fuels and raw materials similar to those produced by petroleum refining, production of hydrogen, alcohols, etc. Thus, its applicability is wide.

RC (e.g., biomass) liquefaction has been studied extensively, but it has never had a wide applicability in industry. This is because it is a process generally done in batches and not continuously, and because it uses unconventional reactors. It is a process that takes place in a liquid medium, with the addition of reduced atmosphere and high temperature. There have been many attempts of application for this process and its variations, and current interest in it is very low.

As for organic synthetic routes (Chapter 2), the reaction control for inorganic and thermal synthetic routes can take place by:

- Thermodynamic control: obtaining the most stable products, which leads to lower values of Gibbs free energy (ΔG) for the reaction
- Kinetic control: obtaining products with lower activation energy (E_{at}) and higher reaction speed

The thermodynamics applied to conversion routes considers the free energy of activation ($\Delta G^{\#}$), and its decrease favors the formation of products (Eq. 3.1):

$$\Delta G^{\#}\left(\mathrm{kJ\,mol^{-1}}\right) = \Delta H^{\#} - T\Delta S^{\#} \tag{3.1}$$

where $\Delta H^{\#}$ is the activation enthalpy ($\mathrm{kJ\,mol^{-1}}$), T is the system temperature (K), and $\Delta S^{\#}$ the activation entropy ($\mathrm{J\,K^{-1}\,mol^{-1}}$).

For the reaction kinetics, in a hypothetical reaction where reactants A and B lead to the formation of product C in the presence of catalyst P, the reaction rate can be expressed as follows (Eq. 3.2):

$$\nu\ \left(\mathrm{s^{-1}\,mol\,L^{-1}}\right) = k[A][B][P] \tag{3.2}$$

where k is the rate constant ($\mathrm{s^{-1}}$, for first-order reactions, or $\mathrm{dm^3\ mol^{-1}\ s^{-1}}$, for second-order reactions), $[A]$ and $[B]$ are the reactant concentrations ($\mathrm{mol\,L^{-1}}$), and $[P]$ is the catalyst concentration ($\mathrm{mol\,L^{-1}}$).

Another way to express the kinetics of a reaction is (Eq. 3.3):

$$\nu\ \left(\mathrm{mol\,L^{-1}\,s^{-1}}\right) = d[C]/\mathrm{dt} \tag{3.3}$$

where $d[C]$ is the concentration of the product formed ($\mathrm{mol\,L^{-1}}$) from its first derivative, and dt is the reaction time (s) also from its first derivative.

However, the reaction rate can be directly proportional to the catalyst concentration (Eq. 3.4):

$$\nu\ \alpha\ [P] \tag{3.4}$$

Therefore, the reaction rate can be inversely proportional to the activation energy of the reaction (Eq. 3.5):

$$\nu\ \alpha\ 1/E_{\mathrm{at}} \tag{3.5}$$

It is convenient highlight the high energetic demands from the thermochemical processes in order to overcome the barrier of activation and to improve the reaction rate.

3.3 Industrial carbon-based inorganic chemicals: A case study

Nanoparticulated carbonates can present interesting properties for use due to surface area and pore volume. For example, their use comprise carriers for drug delivery, pH controllers for acidified soils, and crop production in confined spaces. Fig. 3.5 depicts a process developed to obtain nanocarbonate from CO_2 emitted from fossil fuel (e.g., coal) combustion for energy generation.

The process depicted in Fig. 3.5 is based on Eq. (3.6):

$$\begin{gathered} \mathrm{CO_2\ (g) + Ca(OH)_2\ (aq) \rightarrow CaCO_3\ (s) + H_2O\ (l)}, \\ \Delta_f G^\circ = -1129.1\,\mathrm{kJ\,mol^{-1}} \end{gathered} \tag{3.6}$$

The formation of carbonates (Fig. 3.6) can be a suitable alternative to capture CO_2 from power plants because carbonates of Mg^{2+}, Ca^{2+}, and

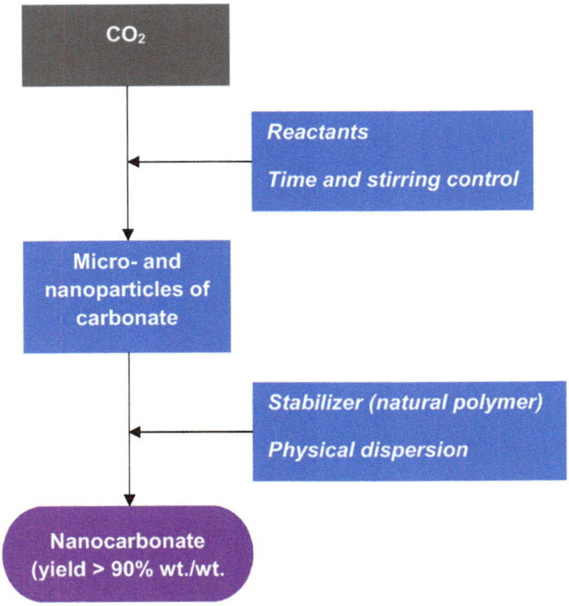

Fig. 3.5 Synthetic inorganic process for nanocarbonate production from CO_2 emitted from fossil energy plants. *(Credit: author.)*

Fig. 3.6 Formation of calcium nanocarbonate (white suspension, A) in a reactor under mild conditions and the dried product (B). *(Credit: author.)*

Ba^{2+} are easily precipitated when CO_2 is added to aqueous hydroxide solutions of these alkaline metals (Vogel et al., 2000). According to the thermodynamic aspects of this reaction, the molecular formation mechanisms of carbonates in aqueous medium involve a prenucleation step for crystallization controlled by system entropy (Kellermeier et al., 2016).

3.4 Industrial carbon-based materials: A case study

Carbon fiber obtained from biomass—for example—can be a source of RC to produce a carbon/carbon composite by means of pyrolysis processing (Fig. 3.7).

Composites are, in a simplified way, two-phase materials with improved physical properties when compared to one-phase materials, e.g., mechanical and temperature resistance, among others. For example, a synthetic composite material can be obtaining by combining a ceramic and a sintered metal creating a combined organic-inorganic matrix. Regarding carbon/carbon composites (Fig. 3.8), major applications involve uses at high temperatures in the aerospace industry.

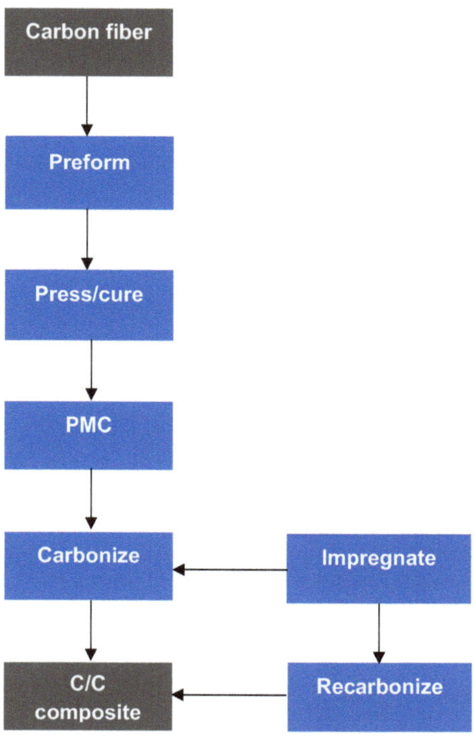

Fig. 3.7 Processing of carbon/carbon composites by pyrolysis of carbon fiber or carbon polymer. *PMC*, processing matrix composite. *(Source: adapted from Chawala, K.K., 2012. Composite Materials – Science and Engineering, third ed. Springer, New York, p. 295. Reproduced with permission from Springer Nature.)*

Fig. 3.8 A carbon/carbon composite material. *(Credit: Shutterstock, https://www.shutterstock.com/pt/image-photo/carbon-fiber-composite-material-background-396357796.)*

3.5 Conclusions

Inorganic and thermochemical routes are useful synthetic strategies to convert RC into valuable chemicals and materials. Thermochemical technologies, despite the high energy demand, can be suitable for biomass sources, whereas inorganic synthesis can be suitable for CO_2.

Products as nanocarbonates, carbon nanotubes, graphene, fullerenes, and composites, among others, should boost the industrial potential of RC raw materials.

References

Azizi-Lalabadi, M., Hashemi, H., Feng, J., Jafari, S.M., 2020. Carbon nanomaterials against pathogens; the antimicrobial activity of carbon nanotubes, graphene/graphene oxide, fullerenes, and their nanocomposites. Adv. Colloid Interface Sci. 284, 102250.

Baby, R., Saifullah, B., Hussein, M.Z., 2019. Carbon nanomaterials for the treatment of heavy metal-contaminated water and environmental remediation. Nanoscale Res. Lett. 14, 341.

Boffetta, P., Jourenkova, N., Gustavsson, P., 1997. Cancer risk from occupational and environmental exposure to polycyclic aromatic hydrocarbons. Cancer Causes Control 8, 444–472.

Bonaventura, D., Chacatergui, R., Valverde, J.M., Becerra, J.A., Verda, V., 2017. Carbon capture and utilization for sodium bicarbonate production assisted by solar thermal power. Energ. Conver. Manage. 149, 860–874.

Bonaventura, D., Chacatergui, R., Valverde, J.M., Becerra, J.A., Ortiz, C., Lizana, J., 2018. Dry carbonate process for CO_2 capture and storage: integration with solar thermal power. Sustain. Energy Rev. 82, 1796–1812.

Idrees, M., Rangari, V., Jeelani, S., 2018. Sustainable packaging waste-derived activated carbon for carbon dioxide capture. J. CO_2 Util. 26, 380–387.

Ji, X., Mu, Y., Liang, J., Jiang, T., Zeng, J., Lin, Z., Lin, Y., Yu, J., 2021. High yield production of 3D graphene powders by thermal chemical vapor deposition and application as highly efficient conductive additive of lithium ion battery electrodes. Carbon 176, 21–30.

Kellermeier, M., Raiteri, P., Berg, J.K., Kempter, A., Gale, J.D., Gebauer, D., 2016. Entropy drives calcium carbonate ion association. ChemPhysChem 17, 3535–3541.

Langanke, J., Wolf, A., Peters, M., 2015. Chapter 5—polymers from CO_2—an industrial perspective. In: Styring, P., Quadrelli, E.A., Armstrong, K. (Eds.), Carbon dioxide utilization: closing the carbon cycle. Elsevier, Amsterdam, pp. 59–71.

Omoriyekomwan, J.E., Tahmasebi, A., Zhang, J., Yu, J., 2019. Mechanistic study on direct synthesis of carbon nanotubes from cellulose by means of microwave pyrolysis. Energ. Conver. Manage. 192, 88–99.

Palmkvist, N.G.M., Nedelec, J.-M., Seisenbaeva, G.A., Kessler, V.G., 2017. Controlling nucleation and growth of nano-$CaCO_3$ via CO_2 sequestration by a calcium alkoxide solution to produce nanocomposites for drug delivery applications. Acta Mater. 57, 426–434.

Petit, P., Maître, A., Persoons, R., Bicout, D.J., 2019. Lung cancer risk assessment for workers exposed to polycyclic aromatic hydrocarbons in various industries. Environ. Int. 124, 109–120.

Vogel, A.I., Mendham, J., Denney, R.C., 2000. Volgel's Quantitative Chemical Analysis, sixth ed. Prentice Hall, New York.

von Broembsen, D., 2007. Apparatus and Method for Producing Activated Carbon. Patent n° US7,288,503B2. United States patent office.

CHAPTER 4

Biochemical synthesis for carbon derivatives

Biochemical routes and processes have been applied to renewable carbon (RC) raw materials for thousands of years for the fermentation of sugars to produce wines and beers. Indeed, these processes were "discovered" and perfected more than 3000 years ago by the Egyptian and Mesopotamian civilizations.

Their uses to promote renewable products is pursued to reduce costs of production despite their limitations, which are discussed in this chapter.

4.1 Introduction to biochemical synthesis applied to renewable carbon products

The transformation of RC sources into different products can occur through the use of chemical, thermochemical, or biochemical processes. The choice of process to be used depends on, among other factors:

- The desired product or the target compound/target material
- Production cost
- Reaction time
- Waste generation
- Raw material availability

The interest in biochemical processes is due to the possibility of converting RC into specific products under mild conditions of temperature, pH value, and pressure when compared against chemical processes; furthermore, it can be related to a sustainability approach based on the use of biological conversion. However, some general limitations of the biochemical processes—mainly microbial processes—are:

- Slow rate of reaction, from hours to days
- Low conversion rate and yield
- Production of a large number of by-products with large quantities of impurities, creating the necessity of more rigorous separation and purification steps

Renewable Carbon
https://doi.org/10.1016/B978-0-323-99735-5.00001-3

- Conversions at aqueous medium for polar compounds (e.g., organic acids), creating the necessity of additional separation (or recovery) step, e.g., by precipitation as a salt

However, some biochemical processes can be economically advantageous, such as the ethanol production by means of yeast fermentation using *Saccharomyces cerevisiae*, which is a well-established industry around the world, e.g., ethanol from sugarcane and corn.

To promote biochemical processes and provide more sustainable chemicals, we have synthetic biology, a modern approach to produce chemicals from biological sources by means of the design and construction of new biological parts and systems, and the redesign of existing, natural biological systems for useful purposes (Nature Portfolio, 2021). It is comprised of the following research, design, and implementation steps (American Chemical Society, 2021):

- Design and optimization of genetic systems
- Genetic circuit design and principles for their organization into programs
- Computational methods to aid the design of genetic systems
- Experimental methods to quantify genetic parts, circuits, and metabolic fluxes
- Genetic parts libraries
- Protein engineering, including computational design
- Metabolic engineering and cellular manufacturing, including biomass conversion
- Natural product access, engineering, and production
- Creative and innovative applications of cellular programming
- Minimal cell design and construction
- Genomics and genome replacement strategies
- Viral engineering
- Automated and robotic assembly platforms for synthetic biology
- DNA synthesis methodologies
- Metagenomics and synthetic metagenomic analysis
- Bioinformatics applied to gene discovery, chemoinformatics, and pathway construction
- Gene optimization
- Methods for genome-scale measurements of transcription and metabolomics
- Systems biology and methods to integrate multiple data sources
- Nucleic acid engineering

Considering biomass as the most prominent RC raw material for biochemical conversion, it is possible to obtain a good number of chemical compounds produced from fossil sources by applying the synthetic biology and immobilized enzymes approaches (discussed later). However, biomass products generally may have a higher production cost than those produced from fossil sources. Therefore, there is a need to improve existing processes and develop new ones that make the production of renewable chemicals economically viable. Currently, some chemical compounds of commercial interest can already be produced from biomass, such as (International Energy Agency, 2020; Biotechnology Innovation Organization, 2018):

- Organic acids: succinic, lactic, malic, citric, propionic, levulinic, aspartic, glucaric, 2,5-furandicarboxylic, and glutamic
- Glycerol and its derivatives: propylene glycol, sorbitol, xylitol, and arabinitol
- Hydrocarbons and ethanol

4.2 Biochemical processes for the synthesis of renewable carbon products

From a practical overview, the biochemical processes can be divided into two main types: *microbial* and *enzymatic.*

Fig. 4.1 depicts the application of these processes for RC raw material conversion into chemicals and biofuels; they are commonly known as bioproducts. We can apply a microbial (A) or enzymatic (B) process depending interest and necessity (product, costs, purity, etc.). After the conversion step, we have a mixture of product(s) and by-product(s) to separate via a downstream step using industrial chromatography, among other technologies. They are then purified and undergo chemical analysis—chromatographic and/or spectroscopic—to confirm the obtained compound (A or B). Generally, immobilized enzymes can be recycled when the microorganisms cannot, except in a few cases such as the yeast *Saccharomyces cerevisiae* for ethanol production.

Microbial processes

In microbial processes, the RC conversion is carried out by fungi, bacteria, and yeasts catalyzed by the action of many enzymes, which can be distributed inside and outside the microorganism (Fig. 4.2). Generally, in addition to the carbon source, other compounds must be added to the bioreactor to

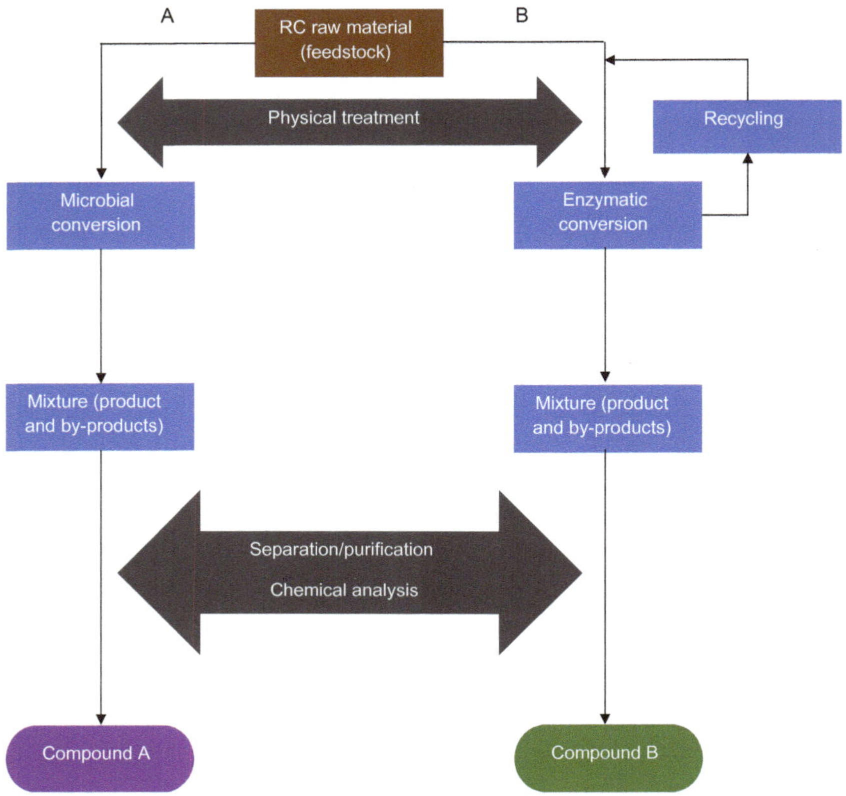

Fig. 4.1 Flowchart of biochemical processes for the application of RC raw material conversion into bioproducts. *(Credit: author.)*

Fig. 4.2 Individual colonies of *Saccharomyces cerevisiae* yeast growing on solid YPD nutrient medium. *(Credit: Shutterstock, https://www.shutterstock.com/pt/image-photo/yeast-under-microscope-475924600.)*

maintain cell viability. Microorganisms can consume different components of RC-generating products through different metabolic routes comprising, for example, aerobic and anaerobic fermentation.

Regarding the lignocellulosic biomass, many microorganisms are able to convert monomers from cellulose and hemicellulose into different compounds, such as alcohols (e.g., butanol, ethanol, propanediol), esters derived from different fatty acid chains (e.g., oleic, lauric, palmitic, myristic, etc.), and small organic acids (e.g., acetic, butyric, citric, fumaric, glycolic, itaconic, kojic, lactic, malic, propionic, succinic, tartaric, and others) (Banner et al., 2021; Bozell and Petersen, 2010).

Many new technologies involving the use of genetically modified microorganisms to produce tailored molecules have emerged in recent years. The improvement in yield for a given product is also one of the research focuses in this area. Companies such as Amyris (United States), Lanzatech (United States), Advanced Enzymes (India), Poet (United States), Reverdia (The Netherlands), Royal DSM (The Netherlands), Stora Enso (Finland), and Myriant (United States) are some examples of biotech companies that have been developing technologies in bioprocesses (Biotechnology Innovation Organization, 2018). Currently, the use of tools such as genomics, proteomics, and bioinformatics provides a large amount of information that facilitates the development of these biochemical processes (de Carvalho et al., 2019).

Biotechnological advances demonstrate that microorganisms can be engineered toward producing plant metabolites as food applications, such as polyphenols and terpenoids, as illustrated in Fig. 4.3; however, precursor supply and product toxicity are key limitations during microbial production (Kallscheuer et al., 2019).

In the case of chemicals with industrial applications, there are some important examples, such as:

- Latic acid from glucose, sucrose, or lactose using *Lactobacillus*, *Sporolactobacillus*, *Enterococcus*, *Lactococcus*, *Bacillus*, *Streptococcus*, and *Pediococcus*;
- Succinic acid from xylose using *Actinobacillus succinogenes* and *Mannheimia succiniciproducens*
- *n*-Butanol from several sugars using engineered *Clostridium butylicum*
- Acetic acid from lignocellulosic biomass using *Lactobacillus casei* and *Clostridium thermoaceticum*

The technical feasibility of using different types of substrates has, in many cases, already been demonstrated. However, the great challenge currently is to make the production of biofuels and chemicals (bioproducts)

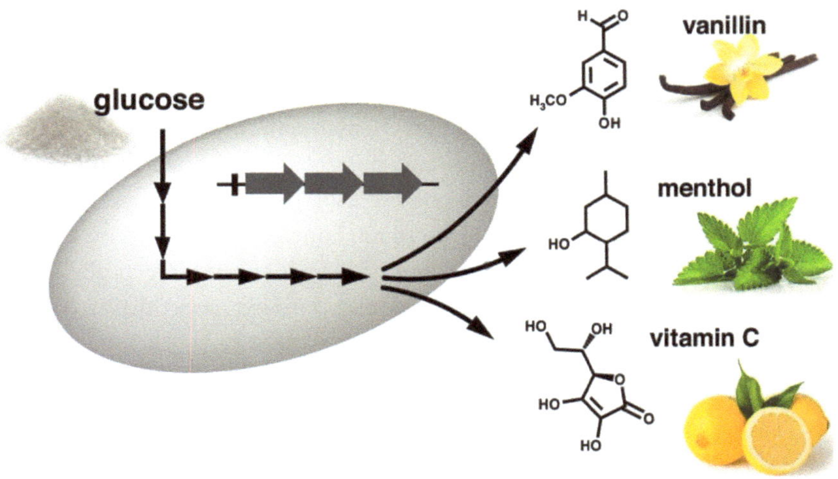

Fig. 4.3 Secondary plant metabolites from glucose produced by engineered microorganisms. *(Credit: Kallscheuer, N., Classen, T., Drepper, T., Marienhagen, J., 2019. Production of plant metabolites with applications in the food industry using engineered microorganisms. Curr. Opin. Biotechnol. 56, 7–17. Reproduced with permission from Elsevier.)*

economically viable. For this, many researchers have sought the development of tailored and direct production processes. An example of this advance is the use of metabolic engineering to produce diesel from cellulose with certain physicochemical properties of interest, through fermentation processes using yeasts and bacteria, or directly from modified photosynthetic algae (Sheldon, 2011).

When using growing microbial cells in a fermentation process, substantial amounts of residual biomass can be generated. Normally, this is directed to animal feed or as a source of energy for the process, hence opportunities for developing new products. However, from these observations, we see some disadvantages of fermentation processes (Koutinas et al., 2007):

- There is a need to use additional nutritional supplement during fermentation.
- The low efficiency in many fermentation processes is due to the use of part of the substrates for cell maintenance and growth, in addition to the formation of other products (by-products) that are not of interest.

Strategies to overcome such limitations involve first performing cell growth and then carrying out the reaction of interest or using isolated enzymes (Ghatak, 2011).

Genetically modified organisms (GMOs), such as bacteria and yeasts, have drawn attention due to the possibility of combining their high potential of producing chemical compounds of different structures in a single fermentation in an optimized and programmed way. Examples of microorganisms that have been modified to increase their tolerance to toxic compounds, metabolize different carbon sources (C5 and C6), and improve product purity and yield are *Escherichia coli*, *Bacillus* sp., lactic acid bacteria, *Corynebacterium glutamicum*, and *Saccharomyces cerevisiae* (Hasunuma et al., 2013).

The modification of microbes that are already used industrially, through genetic or metabolic engineering tools, has taken place in recent years through the advancement of synthetic biology, as previously introduced in Section 4.1. For example, the cyanobacteria used to produce amino acids have been modified to metabolize five-carbon substrates, such as arabinose and xylose, with the advantage of supporting high concentrations of inhibitors such as furfural, hydroxymethylfurfural, and organic acids. The synthesis of other chemical compounds using these microorganisms is also a target of interest (Gopinath et al., 2012).

One of the main problems in biochemical processes is inhibition. For example, the biochemical process of producing succinic acid is inhibited by the product itself, and this limits yield. To solve this problem, continuous removal of the product can be applied, and this can be carried out through precipitation, electrodialysis, and reactive extraction (Lyko et al., 2009) to reach best yields.

Enzymatic processes

Enzymes are proteins and the natural (bio)catalysts that enhance the rate of metabolic reactions. Some enzymes used to convert RC raw materials into renewable chemicals are (Fülöp and Ecker, 2020; Kaushal et al., 2021; Molina-Fernández and Luis, 2021):

- For lignin conversion: lignin peroxidase, manganese peroxidase, and laccase;
- For cellulose conversion: cellulases, e.g., cellobiohydrolase, endoglucanase, and β-glucosidase
- For xylose (from hemicellulose) conversion: xylanases, e.g., endo-1,4-β-xylanase
- For degradation of plastics: hydrolases, e.g., esterases, lipases, depolymerase, and PETases
- For CO_2 capture: carbonic anhydrase

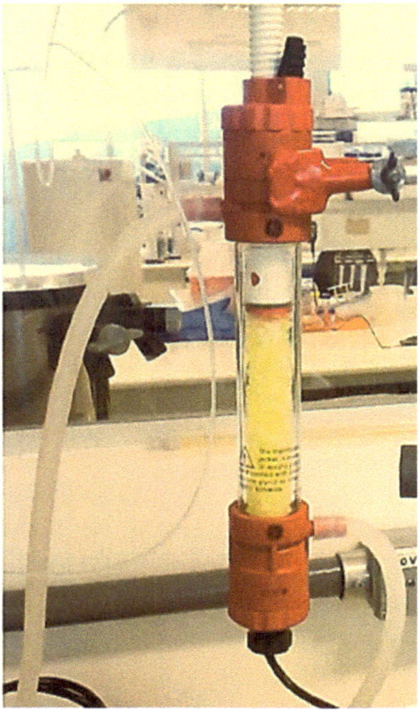

Fig. 4.4 Fixed bed reactor filled with glycosidase enzyme immobilized on alginate beads. *(Credit: author.)*

Immobilized enzymes are, by definition, enzymes retained or confined in a space that can be used repeatedly in a given process (Fig. 4.4). This approach makes use of surface chemistry, pendent functional entities, and ease in tunability of various materials to promote the best enzyme activity and use (Liu et al., 2021). Furthermore, enzyme immobilization has emerged as an important strategy to reduce enzyme costs, as it often enhances enzyme stability while also allowing an easy recovery and reuse (Zanuso et al., 2021). Examples of immobilized enzymes are:

- Membrane reactors, in which the product permeates the membrane, while the enzyme remains retained
- Enzymes linked to insoluble particles (supports), in which at the end of the process it is possible to separate them through filtration

The cell itself can be used as a support for enzyme immobilization because, after cell death, the enzymes can remain actively encapsulated inside the cell, be recovered, and used over and over again.

As for the application of enzymes for the conversion of lignocellulosic biomass components (Fig. 4.5)—the processes involving hydrolysis of

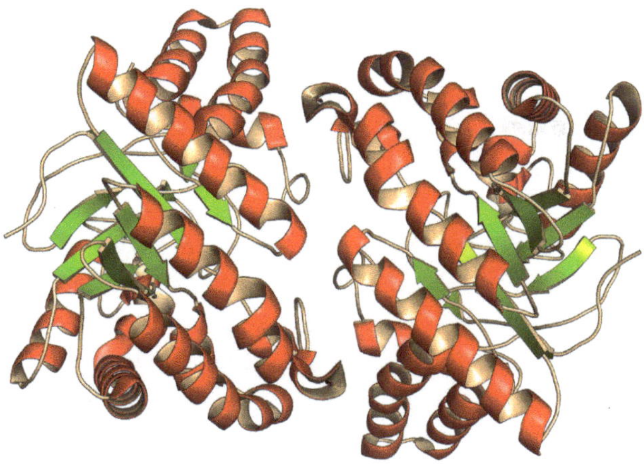

Fig. 4.5 The 3D structure of cellulase (endoglucanase Cel6B from *Humicola insolens*), a cellulose breakdown enzyme. *(Credit: Shutterstock, https://www.shutterstock.com/pt/image-illustration/cellulase-endoglucanase-cel6b-humicola-insolens-cellulose-369113903.)*

starch, cellulose, hemicellulose, and proteins—the esterification of fatty acids and the transesterification of glycerides are currently the most investigated topics.

Despite the advances achieved in the identification and improvement of microorganisms and enzymes for the conversion of sugars from biomass into fuels and chemicals, there is still a great challenge to be overcome in the field of lignocellulosic biomass deconstruction, that is, in the conversion of polysaccharides (cellulose and hemicellulose) into monomers (mainly glucose and xylose) and lignin macromolecules into phenols. Overcoming this hurdle is necessary to reduce competition for sugars from food sources such as corn, beets, etc., and to promote new value chains.

The release of the main monomers present in lignocellulosic biomass is essential for most microorganisms to be able to metabolize the biomass and generate products of interest. Currently, this stage of the process is carried out by chemical or thermochemical processes associated with enzymatic processes. The chemical and thermochemical steps (pretreatment) facilitate the access of enzymes to polysaccharides, promoting the deconstruction of biomass. However, many of these pretreatment processes lead to the degradation of biomass components, generating products that cause inhibition in later biochemical steps. Thus a robust enzymatic cocktail must be designed that could resist inhibitors (Agrawal et al., 2021).

Enzymes are proteins (macromolecules) produced from renewable sources that are biodegradable. They allow a process with less energy consumption and generate less waste. They can catalyze, in a mild and more selective way, synthesis routes with product quality and stability when compared to processes using chemical catalysts. Enzymes are of special interest for green chemistry due to their low environmental toxicity, combined with high selectivity for active sites, which provides a great increase in reaction rate.

Conversions can be performed using isolated enzymes or whole cells. The advantage of using isolated enzymes is that it can be purified, which avoids parallel reactions (such as the action of proteases) in addition to being chemically stabilized. The use of whole cells has the main advantage of not requiring enzyme purification steps. Moreover, in the case of dead cells, the cell structure can act as an immobilization support (Sheldon, 2018).

Many biochemical processes require operationally stable enzymes at high temperatures, as this facilitates mixing, improves substrate solubility, increases mass transfer rate, and reduces the risk of contamination (Turner et al., 2007).

The main problems related to the application of enzymes in industrial processes are their poor stability; they are not easily recycled and work in a limited temperature range. Such limitations can be overcome through the immobilization of enzymes (Gasser et al., 2012).

Thermostable enzymes are generally able to act under conditions required for the process. A thermostable enzyme has a long half-life, even at temperatures above 50°C. Another advantage is that they can be stored at room temperature for a long time, with tolerance to organic solvents, and reduced risk of contamination and loss of activity during their use (Turner et al., 2007).

Enzymes can catalyze a wide range of reactions, e.g., oxidations, reductions, formation of carbon-carbon bonds, and hydrolysis, which are of special interest for obtaining building blocks derived from RC raw materials, such as biomass. Examples of enzymes and their actions include (Turner et al., 2007):

- Monooxygenase, for hydroxylation and oxidation
- Alcohol dehydrogenase, for stereoselective reduction of carbonyl compounds to produce chiral alcohol
- Lipases or oxidoreductases, for epoxide synthesis
- Lipases, for formation of C—C bonds
- Hydrolases and transferases, for synthesis of glycosides

Enzymes with well-established industrial applications are proteases, lipases, and glycosyl hydrolases. On the other hand, three enzymes are required for the conversion of xylan to xylitol: xylanase, xylosidase, and xylose reductase. However, the conversion of xylose to xylitol depends on the use of cofactors, such as NADPH, to reduce the cost of the process; the use of an enzyme that recycles the cofactor or uses the whole cell is necessary. Some yeasts can produce xylitol; however, unwanted reactions such as xylulose production are observed (Turner et al., 2007).

An interesting research topic is the development of multifunctional catalysts. The idea is to bring together different active sites in the same structure, or catalytic particle, improving the reaction rate and yield of the desired product. Enzymatic and inorganic catalysts can be combined in this system, as robust and easily regenerated catalysts must be developed to enable their application in complex media, containing many impurities (Gallezot, 2012), which is characteristic of biomass components.

Regarding the kinetic aspects of the enzymatic conversions, it is governed by the Michaelis-Menten equation (Eq. 4.1):

$$E + S \underset{k_{-1}}{\overset{k_1}{\rightleftharpoons}} ES \overset{k_2}{\rightarrow} E + P \tag{4.1}$$

where ES is the enzyme-substrate complex, E is the enzyme, S is the substrate, P is the product, and k_n the rate constant.

The derivation of Eq. (4.1) for the production rate of a certain product P is:

$$\nu_{\max} = k_{\text{cat}} [E_0] \tag{4.2}$$

It is worth considering that, in general, an enzyme catalyzes only one reaction type (reaction specifity) and operates on only one type of substrate (substrate specifity). Substrate molecules are attacked at the same site (regiospecifity), and only one or preferentially one of the enantiomers of chiral substrates or racemic mixtures is attacked (stereospecifity) (International Union of Pure and Applied Chemistry, 1992).

4.3 Industrial carbon-based biochemical: A case study

Succinic acid (Fig. 4.6) corresponds to a well-recognized platform chemical utilized as a precursor in industrial synthesis of various chemical compounds as green plastics. It is an intermediate metabolite in the citric acid cycle (Fig. 4.7). It has a role as a nutraceutical, a radiation protective agent, an anti-ulcer drug, a micronutrient, and a fundamental metabolite.

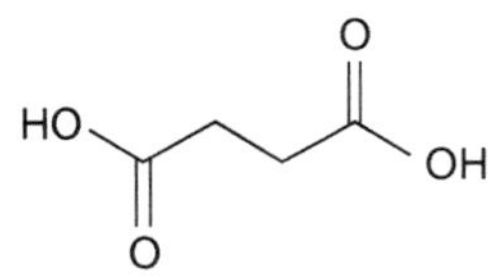

Fig. 4.6 Succinic acid structure. *(Credit: author.)*

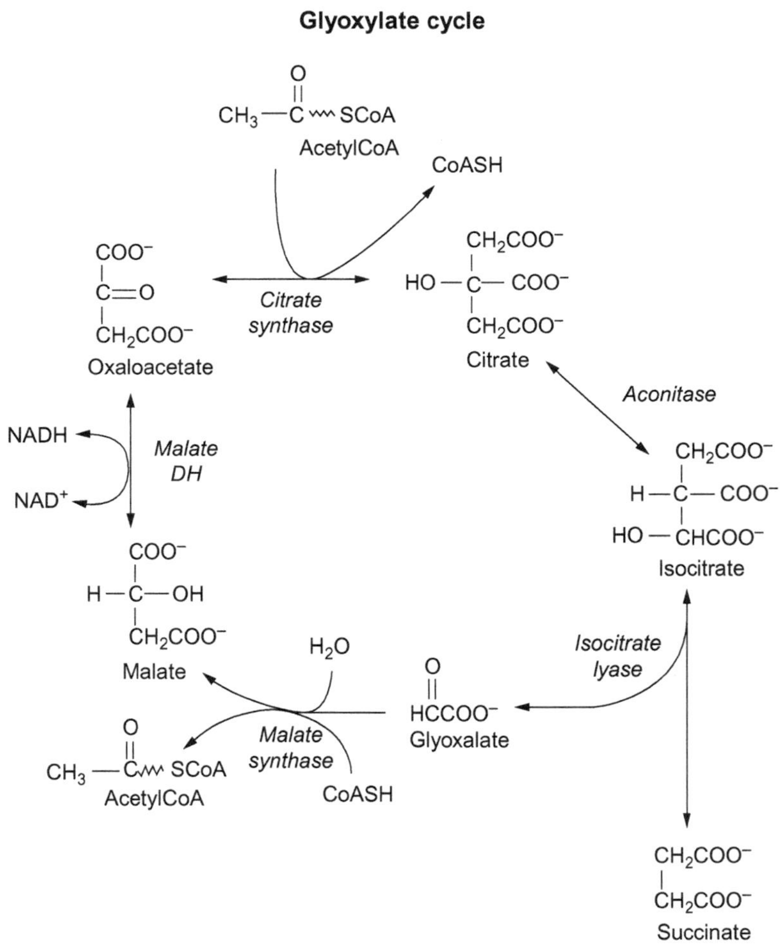

Fig. 4.7 Metabolic route for succinic acid production from the glyoxylate cycle pathway. *Credit: Goldberg, I., Rokem, J.S., 2009. Organic and fatty acid production, microbial. In: Schaechter, M. (Ed.) Encyclopedia of Microbiology. third ed. Academic Press, Cambridge, pp. 421–442. Reproduced with permission from Elsevier.*

Succinic acid can be produced through fermentation of sugars from different biomass, such as starch or sugar crops, and lignocellulosic materials using well-known industrial microorganisms (such as *Escherichia coli* or *Saccharomyces cervisiae*) and modifying their minor succinic acid production capability into high yields through metabolic engineering (European Commission, 2021). The process discussed here allows for the production of succinic acid from the C5 fraction (xylose from hemicellulosic fraction) of sugarcane bagasse, which is its main differential because the common source of RC is the C6 (glucose).

The production of succinic acid and its anhydride occurs, preferably, by catalytic hydrogenation of maleic acid and its anhydride or by its recovery as a by-product of adipic acid synthesis (Fumagalli, 2006). However, different companies around the world developed biotechnological routes for the production of succinic acid, such as:

- BioAmber (Canada)
- Corbion (The Netherlands)
- Myriant (United States)
- Reverdia (The Netherlands)

Generally, they use fermentation by yeast as the industrial microbial process.

Base scenarios consist of consolidated or promising industrial routes for the production of hemicellulose derivatives studied in the project. Their simulation, in the Sugarcane Virtual Biorefinery,[a] allow for comparison of the current production routes with the routes under development to understand advantages and disadvantages. Because it is a route that is already fully consolidated, it was decided to use the petrochemical route as the base scenario. As the routes under development use renewable sources, analysis of the petrochemical routes allows for their comparison.

Scenarios can consist of routes that use the same starting material (e.g., sugarcane bagasse) used by the routes under development or routes that use alternative raw materials, such as different lignocellulosic residues or petrochemical derivatives.

There are a few different scenarios for obtaining succinic acid and its anhydride via the petrochemical route. Patents have been found describing the use of both maleic anhydride and maleic acid as a starting substance (Fumagalli, 2006; Trunin et al., 2013; Darsow, 1999).

From these previous considerations, Klein et al. (2017) performed technical-economic evaluation for succinic acid production from

[a] https://lnbr.cnpem.br/en/facilities/virtual-sugarcane-biorefinery/.

lignocellulosic biomass. The use of hemicellulosic hydrolysate from sugarcane bagasse allows production of succinic acid without competing with food chains and increasing the biomass-aggregated value. This work assessed the economic impacts of integrating the production of succinic acid on an optimized sugarcane biorefinery with the production of first-generation ethanol and electricity. Process simulation was performed to obtain mass and energy balances of the entire process, from bagasse pretreatment to pentose fermentation and succinic acid recovery, using several pieces of technical information and data taken from the scientific literature. After assessing the impacts of key economic variables through risk analysis, the obtained results were benchmarked against those of an optimized ethanol distillery. Calculated succinic acid production cost was US \$2.32 kg^{-1}, comparable to costs related to less expensive sugar sources. The integrated succinic acid biorefinery presented a slightly lower internal rate of return (IRR) compared to an ethanol distillery. However, even when considering uncertainties, succinic acid production has a strong probability (84.6%) of achieving an IRR higher than 12%.

4.4 Conclusions

Despite their limitations when compared against chemical routes and processes—e.g., slow reaction rate, low conversion rate and yield, production of a large number of by-products with large quantities of impurities, and conversions at aqueous medium for polar compounds—biochemical processes have contributed to explore the whole potential of RC raw materials, especially biomass.

The production of several compounds by biochemical processes such as organic acids, alcohols, and esters, among others, can contribute to increase the establishment of renewable bioproducts, i.e., chemicals and biofuels.

References

Agrawal, R., Verna, A., Singhania, R.R., Varjani, S., Dong, C.D., Patel, A.K., 2021. Current understanding of the inhibition factors and their mechanism of action for the lignocellulosic biomass hydrolysis. Bioresour. Technol. 332, 125042.

American Chemical Society, 2021. ACS Synthetic Biology. https://pubs.acs.org/page/asbcd6/about.html. Accessed December 2021.

Banner, A., Toogood, H.S., Scrutton, N.S., 2021. Consolidated bioprocessing: synthetic biology routes to fuels and fine chemicals. Microorganisms 9, 1079.

Biotechnology Innovation Organization, 2018. Renewable Chemical Platforms Building the Biobased Economy. http://go.bio.org/rs/490-EHZ-999/images/BIO_Chemical_Companies_Report_2018_FINAL.pdf. accessed December 2021.
Bozell, J.J., Petersen, G.R., 2010. Technology development for the production of biobased products from biorefinery carbohydrates—the US Department of Energy's "top 10" revisited. Green Chem. 12, 539–554.
Darsow, G., 1999. Process for preparing succinic anhydride. Patent US 5, 952, 514.
de Carvalho, L.M., Borelli, G., Camargo, A.P., de Assis, M.A., de Ferraz, S.M.F., Fiamenghi, M.B., José, J., Mofatto, L.S., Nagamatsu, S.T., Persinoti, G.F., Silva, N.V., Vasconcelos, A.A., Pereira, G.A.G., Carazzolle, M.F., 2019. Bioinformatics applied to biotechnology: a review towards bioenergy research. Biomass Bioenergy 123, 195–224.
European Commission, 2021. Environmental Factsheet: Succinic Acid. https://ec.europa.eu/jrc/sites/default/files/BISO-EnvSust-Bioproducts-Succinic-Acid_141031.pdf. accessed December 2021.
Fülöp, L., Ecker, J., 2020. An overview of biomass conversion: exploring new opportunities. PeerJ Life Environ. 8, e9586.
Fumagalli, C., 2006. Succinic acid and succinic anhydride. In: Kirk, R.E., Othmer, D.E. (Eds.), Kirk-Othmer Encyclopedia of Chemical Technology, third ed. Wiley-Interscience, New York.
Gallezot, P., 2012. Conversion of biomass to selected chemical products. Chem. Soc. Rev. 41, 1538–1558.
Gasser, C.A., Hommes, G., Schäffer, A., Corvini, P.F.-X., 2012. Multi-catalysis reactions: new prospects and challenges of biotechnology to valorize lignin. Appl. Microbiol. Biotechnol. 95, 1115–1134.
Ghatak, H.R., 2011. Biorefineries from the perspective of sustainability: feedstocks, products, and processes. Renew. Sustain. Energy Rev. 15, 4042–4052.
Gopinath, V., Murali, A., Dhar, K.S., Nampoothiri, K.M., 2012. *Corynebacterium glutamicum* as a potent biocatalyst for the bioconversion of pentose sugars to value-added products. Appl. Microbiol. Biotechnol. 93, 95–106.
Hasunuma, T., Okazaki, F., Okai, N., Hara, K.Y., Ishii, J., Kondo, A.A., 2013. Review of enzymes and microbes for lignocellulosic biorefinery and the possibility of their application to consolidated bioprocessing technology. Bioresour. Technol. 135, 513–522.
International Energy Agency, 2020. Bio-Based Chemicals—A 2020 Update. https://www.ieabioenergy.com/wp-content/uploads/2020/02/Bio-based-chemicals-a-2020-update-final-200213.pdf. accessed December 2021.
International Union of Pure and Applied Chemistry, 1992. Glossary for chemists of terms used in biotechnology. Pure Appl. Chem. 64, 143–168.
Kallscheuer, N., Classen, T., Drepper, T., Marienhagen, J., 2019. Production of plant metabolites with applications in the food industry using engineered microorganisms. Curr. Opin. Biotechnol. 56, 7–17.
Kaushal, J., Khatri, M., Arya, S.K., 2021. Recent insight into enzymatic degradation of plastics prevalent in the environment: a mini—review. Cleaner Eng Technol 2, 100083.
Klein, B.C., Silva, J.F.L., Junqueira, T.L., Rabelo, S.C., Arruda, P.V., Ienczak, J.L., Mantellato, P.E., Pradella, J.G.C., Junior, S.V., Bonomi, A., 2017. Process development and techno-economic analysis of bio-based succinic acid derived from pentoses integrated to a sugarcane biorefinery. Biofuels Bioprod. Biorefin. 11, 1051–1064.
Koutinas, A.A., Arifeen, N., Wang, R., Webb, C., 2007. Cereal-based biorefinery development: integrated enzyme production for cereal flour hydrolysis. Biotechnol. Bioeng. 97, 61–72.
Liu, S., Bilal, M., Rizwan, K., Gul, I., Rashed, T., Iqbal, H.M.N., 2021. Smart chemistry of enzyme immobilization using various support matrices—a review. Int. J. Biol. Macromol. 190, 396–408.

Lyko, H., Deerberg, G., Weidner, E., 2009. Coupled production in biorefineries — combined use of biomass as a source of energy, fuels and materials. J. Biotechnol. 142, 78–86.

Molina-Fernández, C., Luis, P., 2021. Immobilization of carbonic anhydrase for CO_2 capture and its industrial implementation: a review. J. CO_2 Util. 47, 101475.

Nature Portfolio, 2021. Synthetic Biology. https://www.nature.com/subjects/synthetic-biology. accessed December 2021.

Sheldon, R.A., 2011. Utilisation of biomass for sustainable fuels and chemicals: molecules, methods and metrics. Catal Today 167, 3–13.

Sheldon, R.A., 2018. Enzymatic conversion of first- and second- generation sugars. In: Vaz Jr., S. (Ed.), Biomass and Green Chemistry—Building a Renewable Pathway. Springer Nature, Cham, pp. 169–183.

Trunin, R.A., Uchitel, M.L., Mevsky, E.I., Heifets, V.I., Kasseinova, D., Pivonenkova, L.P., Lubimova, T.B., 2013. Method for preparing succinic acid. Patent US 8,546,611B2.

Turner, P., Mamo, G., Karlsson, E.N., 2007. Potential and utilization of thermophiles and thermostable enzymes in biorefining. Microb. Cell Fact. 6, 1–23.

Zanuso, E., Gomes, D.G., Ruiz, H.A., Teixeira, J.A., Domingues, L., 2021. Enzyme immobilization as a strategy towards efficient and sustainable lignocellulosic biomass conversion into chemicals and biofuels: current status and perspectives. Sustainable Energy Fuels 5, 4233–4247.

CHAPTER 5

Conclusions and remarks of the scientific aspects for renewable carbon

Remarks and conclusions are approaches used to facilitate the book's understanding and its applicability in real situations. Remarks cover the most representative statements described in the text while conclusions critically summarize the conducted discussions. Both can then be used as a practical compilation, which take parts of relevant contents from each chapter. This didactic approach comprises:

- The reading of the text
- The understanding of its content
- The application of its fundamentals to the renewable carbon (RC) concept related to the scientific basis offered by Part 1

5.1 Remarks

We saw in Chapter 1 that the circular economy, an economic strategy to reach the renewability, is defined by the European Parliament (2021) as "*a model of production and consumption that involves sharing, leasing, reusing, repairing, reforming, and recycling existing materials and products for as long as possible.*" In this way, the life cycle of the products is extended. Moreover, according to the Organisation for Economic Co-operation and Development (2018), a "*bioeconomy refers to the set of economic activities related to the invention, development, production, and use of biological products and processes.*" That said, it is easy to see that the bioeconomy has a broad scope, and its applicability to sustainable products is obvious. From these statements, the *renewable carbon strategy* based on the reuse of carbon sources can be a valuable contribution to rethink the established economic model to promote the circular economy and bioeconomy, because it promotes carbon dioxide and recyclable materials as industrial sources of carbonaceous raw material biomass (Renewable Carbon Initiative, 2021).

https://doi.org/10.1016/B978-0-323-99735-5.00008-6

We saw in Chapter 2 that carbon-derived products are the basis of modern society. For example, the polymer industry is one of the most advanced sectors since the development of polymer synthesis of plastics from oil. Moreover, it has expanded to several classes of molecules (e.g., acids, esters, alcohols, fatty acids, hydrocarbons, etc.) and applications, such as pharmaceuticals, agrochemicals, fuels, food, feed, and materials, among others. As observed by Kühlborn et al. (2020), along with the utilization of biomass for fuel and energy production, the search for new sources of chemical feedstocks is a growing field in current chemical research. Numerous recent publications deal with the production of low molecular weight compounds from biomass and the valorization of these renewable feedstocks for synthetic purposes, although commodities such as adipic acid, acrylic acid, or butanol currently still dominate the field. Furthermore, since organic synthesis is about constructing carbon skeletons, the origin and fate of carbon atoms significantly contributes to the sustainability of a certain organic product. Thus, the renewability of raw materials, i.e., precursors, is a major aspect of green chemistry attended by the RC sources. Therefore, the use of biomass to promote green chemistry, by means of synthetic routes, will generate new products, processes, and value chains as alternatives to fossil raw materials (Vaz, 2018). Moreover, it can be applied to CO_2-derived products. Nevertheless, it is very limited or impracticable for recyclable materials due their physical properties (e.g., physical state) and impurities or contaminants added during the processing step.

We saw in Chapter 3 that, as well as organic synthesis, inorganic synthesis has a purpose in the conversion of a certain raw material (precursor) into a final product (target compound). As opposed to organic synthesis, which can be applied to biomass and carbon dioxide, inorganic synthesis is restricted to carbon dioxide as a RC raw material. It occurs because CO_2 is an inorganic molecule generating inorganic chemicals and materials as final products. However, the thermochemical approach can generate routes of synthesis for some interesting materials from biomass and renewable materials (e.g., industrial polymers). Thermochemical technologies are very common for petrochemical and biomass industries. In some cases, it can be applied to degrade plastic vessels (recyclable materials) contaminated by chemical residues, such as pesticides, as a treatment technology; nevertheless, these technologies are high-energy dependent with increasing costs and without a direct contribution to green chemistry, i.e., the principle of energy economy.

Finally, we saw in Chapter 4 that interest in the biochemical processes is due to the possibility of converting RC into specific products under mild conditions of temperature, pH value, and pressure—with a sustainability approach—when compared against chemical processes. Considering biomass as the most prominent RC raw material, it is possible to obtain a good number of chemical compounds produced from fossil sources by applying synthetic biology allied to immobilized enzymes. However, biomass products generally have a higher production cost than those produced from fossil sources. Therefore, there is a need to improve existing processes and develop new ones that make the production of renewable chemicals economically viable. Furthermore, from a practical overview, the biochemical processes can be divided into two main types: microbial and enzymatic.

From these statements, we can observe that RC strategy demands a considerable number of conversion processes based on the concepts of circular economy, bioeconomy, and green chemistry.

5.2 Conclusions

We concluded in Chapter 1 that the definition of RC is comprised of three components, biomass, carbon dioxide, and recyclable materials, which are largely produced and used by modern society. Indeed, biomass, carbon dioxide, and recycled plastics are potential raw materials for several products, such as chemicals, materials, and pharmaceuticals, among others. The understanding of these three sources of RC and their physicochemical properties can start the development of sustainable value chains to substitute for fossil raw materials (considering biomass and carbon dioxide) or to reuse those yet produced (considering the recyclable materials). Moreover, the carbon atom is one of the most versatile constituents of organic molecules acting as the skeleton for derived products and a perfect target for the application of green chemistry principles.

We concluded in Chapter 2 that organic synthesis presents a huge contribution to reach the RC concept. Its reactions are paramount to obtain renewable products, i.e., chemicals and materials, because most of these products are organic molecules that need a conversion step applied to the feedstock materials. Biomass and carbon dioxide are the main sources of RC for organic synthesis; recyclable materials generally need a physical or thermochemical process for their reuse. The heterogeneity of the RC raw materials, especially for biomass, point to a necessity for a combination of organic reactions to reach the best exploitation of these feedstocks.

We concluded in Chapter 3 that inorganic and thermochemical routes are useful synthetic strategies to convert RC into valuable chemicals and materials. Thermochemical technologies, despite the high-energy demand, can be suitable for biomass sources, whereas direct inorganic synthesis can be suitable for CO_2. Products such as nanocarbonates, carbon nanotubes, graphene, fullerenes, and composites, among others, should boost the industrial potential of the RC raw materials.

Finally, we concluded in Chapter 4 that, despite their limitations when compared against chemical routes and processes—e.g., slow reaction rate, low conversion rate and yield, production of many by-products with large quantities of impurities, and conversions of aqueous medium for polar compounds—biochemical processes have contributed a lot to explore the whole potential of RC raw materials, especially for biomass. The production of several compounds by biochemical processes such as organic acids, alcohols, and esters, among others, can contribute to the establishment of renewable bioproducts, i.e., chemicals and biofuels.

RC has, indeed, a huge potential to become one the most significant sources of industrial carbonaceous raw materials.

References

European Parliament, 2021. Circular Economy. https://www.europarl.europa.eu/thinktank/infographics/circulareconomy/public/index.html. (Access December 2021).

Kühlborn, J., Groβ, J., Opatz, T., 2020. Making natural products from renewable feedstocks: back to the roots? Nat. Prod. Rep. 37, 380–424.

Organisation for Economic Co-operation and Development, 2018. Meeting Policy Challenges for a Sustainable Bioeconomy. https://www.oecd.org/publications/policy-challenges-facing-a-sustainable-bioeconomy-9789264292345-en.htm. (Access December 2021).

Renewable Carbon Initiative, 2021. The Renewable Carbon Initiative. https://renewable-carbon-initiative.com/. (Access December 2021).

Vaz Jr., S. (Ed.), 2018. Biomass and Green Chemistry—Building a Renewable Pathway. Springer Nature, Cham.

PART 2

Technology

CHAPTER 6

Introduction to technological approaches for renewable carbon

Technological tools allow the conversion of RC raw materials into products with greater added value, such as chemicals, materials, and fuels, among others.

These tools are derived from chemical, biochemical, and thermochemical routes and are introduced in this chapter to prepare for detailed discussions in Part 2.

6.1 Main industrial sources of renewable carbon

As introduced in Part 1, RC raw materials can be obtained from three main sources:

- Biomass
- CO_2
- Recyclable materials

These three categories of RC sources are all-encompassing and present in the daily life of modern society, from the production of goods to their consumption.

Considering the industrial sources of these RC materials, it can be related to the following industrial sectors:

- Agriculture and agroindustry for biomass, their products, and residues
- Energy generation from fossil fuels, chemical industry, iron, steel, and cement for CO_2
- Plastics (in a general way)—comprised of resins and synthetic fibers—for recyclable materials

These industrial sectors (Fig. 6.1) are well established worldwide with several value chains with huge productions, as depicted in Table 6.1. However, they can promote negative environmental impacts which can be mitigated by means the technologies presented here.

Although it's not possible to establish a direct relationship among the values presented in Table 6.1, it is possible to note the highlighted contribution by CO_2, a greenhouse gas (GHG), to the availability of RC; that means conversion technologies for this gas have higher relevance as mitigating strategies.

Renewable Carbon
https://doi.org/10.1016/B978-0-323-99735-5.00013-X

Fig. 6.1 Examples of RC raw materials production. From left to right and from top to bottom: (A) agricultural production (source of biomass); (B) coal mining for energy generation (source of CO_2); (C) cement plant (source of CO_2); (D) oil and gas refinery (source of fossil fuels, chemicals and materials, and CO_2); (E) iron and steel plant (source of CO_2); and (F) plastic fabrication (source of recyclable material). *(Credit: https://www.shutterstock.com/.)*

An interesting case in CO_2 generation is the chemical industry, as the gas is generated from processing, fuel, and power. Direct CO_2 emissions from primary chemical production were 880 $MtCO_2$ in 2018 (International Energy Agency, 2021). Despite being the largest industrial energy consumer, it is the third industry subsector in terms of direct CO_2 emissions behind cement, iron, and steel industries. This is largely because around half

Table 6.1 Global production of the main industrial sectors related to the renewable carbon raw material sources.

Sector	Global production (by reference year)	Associated renewable carbon raw material
Agriculture and agroindustry	7.26 Gt (2018)	Biomass
Energy from fossil fuels (coal, oil, and gas)	136,761 TWh (2019)	CO_2
Cement	4.1 Gt (2020)	CO_2
Chemical industry (excluding pharmaceuticals)	946 Mt of transported chemicals produced in the United States (2020)[a]	CO_2
Iron	2.4 Gt (2020)	CO_2
Plastics (including polymer resins and synthetic fibers)	368 Mt (2019)	Recyclable materials
Steel	1.86 Gt as crude steel (2020)	CO_2

Compilation from the most recent data available from American Chemistry Council (https://www.americanchemistry.com/chemistry-in-america/data-industry-statistics/resources/2021-guide-to-the-business-of-chemistry), International Energy Agency (https://www.iea.org/fuels-and-technologies), United Nations Food and Agriculture Organization (https://www.fao.org/faostat/en/), United Nations Industrial Development Organization (https://www.unido.org/researchers/statistical-databases), World Bank (https://data.worldbank.org/), and World Trade Organization (https://data.wto.org/en).

[a]Absence of concise data of global production.

of the chemical subsector's energy input is consumed as feedstock, i.e., fuel (basically hydrocarbons) used as raw material input rather than as a source of energy (International Energy Agency, 2021).

Regarding agriculture and agroindustry sectors, the 7.26 Gt of total production derived from biomass categories (seen in Chapter 1) as presented in Table 6.1 generates an estimative of 140 Gt of dry mass waste, highlighting the lignocellulosic biomass for this value (Vaz Jr., 2020).

Plastics—including polymer resins and synthetic fibers—are a huge source of recyclable materials with a consequent and well-defined impact on the environment and its ecosystems. Thereby technologies for their recycling represent the core for their sustainable use.

Fig. 6.2 depicts a general outlook on the carbon-derived market.

From Fig. 6.2, it is possible to observe a major demand (55%) from the production of chemicals and materials, probably because they constitute the base for several other industry sectors, as civil construction, medicines, and consumer goods, among others.

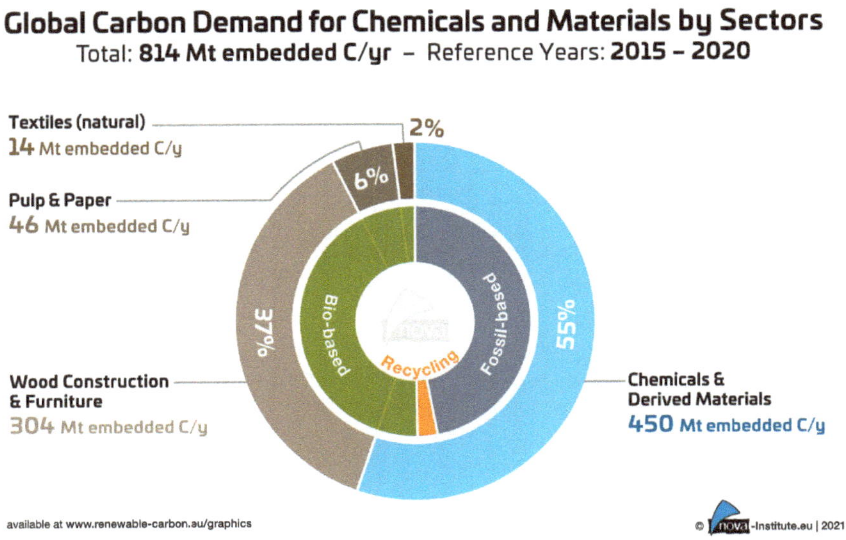

Fig. 6.2 Global carbon-derived products for the reference period of 2015–20. *(Credit: Porc, O., Carus, M., Kähler, F., vom Berg, C., 2021. Global Carbon Demand for Chemicals and Materials by Sectors. https://renewable-carbon.eu/publications/product/global-carbon-demand-for-chemicals-and-materials-by-sectors-png/, accessed December 2021. Reprinted with permission from nova-Institute.)*

6.2 Main technologies for renewable carbon conversion and their challenges

As also introduced in Part 1, there are categories of conversion routes for RC raw sources, which make use of their respective technologies and processes. These routes are divided into:

✓ Biochemical routes, based on enzymes and microbes—especially useful for biomass

✓ Organic routes, based on reactions for compounds of carbon, oxygen, hydrogen, and nitrogen—especially useful for biomass and for CO_2 (with less possibility of application)

✓ Inorganic routes, based on reactions for inorganic compounds—especially useful for CO_2

✓ Thermochemical routes, based on degradation and posterior formation of new products using thermal energy—especially useful for biomass and recyclable materials

Table 6.2 describes the main technologies and examples of applications for RC raw material conversion and their current status.

Additional to those technologies presented in Table 6.1, we can cite the mechanical recycling process for recycling plastics, where postconsumer plastics are sorted, washed, remelted, and extruded (Gu et al., 2017). However, the quality of the obtained products is poor, and it cannot be applied to thermoset plastics. Due to its limitation, this type of process is not a subject of interest in this book.

Regarding challenges to establish technologies for RC raw materials conversion as those described in Table 6.2, they could rise from:

Table 6.2 The current status of technologies for conversion of RC raw materials into valuable products.

Technology	Route	Status	References
Immobilized enzymes for clean production of reducing sugar from lignocellulosic biomass	Biochemical	Under development and demonstration	Lin et al. (2020)
Synthetic biology for microbial lignin biotransformation	Biochemical	Under research and development	Yagushi et al. (2021)
Batch synthesis with or without catalysts for sugar derivatives	Organic	Commercially available and under development	Sanborn (2019)
Flow synthesis for biodiesel	Organic	Under research and development	Tran et al. (2017)
Batch synthesis for mineralization	Inorganic	Under development and at demonstration	Liu et al. (2021)
Combustion for lignocellulosic biomass	Thermochemical	Commercially available and under development	Yuan et al. (2019)
Gasification of lignocellulosic biomass	Thermochemical	Under development and at demonstration	Henning and Haase (2021)
Pyrolysis for cracking of polyethylene waste	Thermochemical	Under development	Cheng et al. (2021)

- Difficulty to establish RC sources as competitive industrial feedstocks as an alternative to oil, which reduces the opportunity for conversion technologies related to these sources
- High costs of industrial plants dedicated to RC, which tends to be reduced with increasing production scale
- Scarce commercial availability of dedicated catalysts for biomass and CO_2 conversion, mainly for chemical and thermochemical processes
- Absence or little of governmental support for renewable or green product development, market insertion, and commercialization

However, these same challenges can turn into opportunities for the RC approach (seen in the Chapter 1) because they can boost the development of new technologies and value chains, and spread wealth in the society.

Additionally, the scientific challenges involve the creation of academic and industrial environments conducive to the development of specialized labor, which involves training and qualification at the technical, undergraduate, graduate, specialization, master, doctoral, and postdoctoral levels. These aspects require a strategic vision of the public and private sectors, with a constant partnership between both stakeholders.

Another scientific challenge to be overcome concerns the development of local knowledge and not just the import of technologies, as is often the case in developing countries. This is an issue that defines a country as an effective player in the scenario; as a secondary country from the point of view of high value, and know-how; or one that only produces and exports raw materials and commodities only to buy them later as consumer goods.

6.3 Technology readiness level assessment

Methodologies to rank technologies according to their readiness are very useful to understand scenarios and define the best investment of time and resources. Also, monitoring the development of these new technologies and the optimization of existing ones are important as they enable future projections for a particular branch of research, development, and implementation (RDI).

Technological maturity assessment is a tool developed by NASA in 1990 that was adapted by several institutions and industries around the world such as the Technology Readiness Level (TRL) system, which is capable of leveling existing technologies and pointing out the status in which they are found (Freeman and Bhown, 2011).

TRL has a simple nomenclature (TRL1, TRL2, TRL3, TRL4, TRL5, TRL6, TRL7, TRL8, and TRL9) and allows analyzing the development of technologies through stages. These stages progress gradually, and each one of them is a prerequisite for the next, as shown in Table 6.3.

From the TRL assessment in Table 6.3, the technologies presented in Table 6.2 can be ranked as follows:

✓ Immobilized enzymes for clean production of reducing sugar from lignocellulosic biomass: TRL3 to TRL8
✓ Synthetic biology for microbial lignin biotransformation: TRL1 to TRL5
✓ Batch synthesis with or without catalysts for sugar derivatives: TRL3 to TRL9
✓ Flow synthesis for biodiesel: TRL1 to TRL5
✓ Batch synthesis for mineralization: TRL3 to TRL8
✓ Combustion for lignocellulosic biomass: TRL3 to TRL9
✓ Gasification of lignocellulosic biomass: TRL3 to TRL8
✓ Pyrolysis for cracking of polyethylene waste: TRL3 to TRL5

The assessment through TRL stages allows projections of time, cost, and environmental impacts; quantifies existing technologies; and helps decision making. Currently, it is a tool used worldwide by political leaders, scientists, and entrepreneurs in various fields, especially in RDI associated with industrial processes and products.

6.4 Advantages and disadvantages of renewable carbon feedstock

The RC approach has its pros and cons. Its advantages and disadvantages need to be critically evaluated through the lens of science, technology, and economy.

The main advantages for the use of RC as industrial feedstock are:

✓ Sustainability characteristics assigned to processes and products
✓ Transformation of wastes into products of high value or with more value than the original raw material
✓ Ecofriendly value chains
✓ Increasing in the equality of wealth for society

Unfortunately, the use of RC as industrial feedstock has some disadvantages, such as:

✓ Higher initial costs for both processes and final products, compared to some petrochemicals

Table 6.3 General TRL assessment and ranking for technology evaluation.

Technology maturity level	Main characteristics	Stage
9	Normal trade service	Commercialization
8	Commercial demonstration, large-scale deployment in final form	Demonstration
7	Subscale demonstration, full functional prototype	
6	Fully integrated pilot testing in a relevant environment	
5	Subsystem validation in a representative environment	Development
4	System validation in a laboratory environment	
3	Proof of concept tests, component level	
2	Application formulation	Research
1	Observation of basic principles, initial concept	

Source: adapted from Freeman, B.C., Bhown, A.S., 2011. Assessment of the technology readiness of post-combustion CO_2 capture technologies. Energy Procedia 4, 1791–1796.

✓ Possible difficulties in the supply of raw materials regarding to quantity and frequency

Despite some disadvantages, the RC advantages as feedstock are more consistent and can contribute to the sustainable development of modern society. Indeed, the model of circular economy introduced in Chapter 1 is only feasible to apply through technologies that enable the use of RC global stock.

6.5 Conclusions

RC sources comprise huge value products and value chains, as depicted by the sectors of agriculture and agroindustry, energy generation, chemical industry, iron, steel, cement, plastics (in a general way), resins, and synthetic fibers.

The use of RC raw materials (CO_2, biomass, and recyclable materials) depends on the application of technologies related to biochemical, organic, inorganic, and thermochemical synthetic routes, which can be evaluated by means the TRL assessment approach. However, there are observed challenges related to science and economy, among other aspects, to be considered for technological advancement.

Finally, the advantages of RC as industrial feedstock overcomes its disadvantages for sustainable development.

References

Cheng, L., Zhang, Y., Wang, Y., Gu, J., Yuan, H., Chen, Y., 2021. Pyrolysis of long chain hydrocarbon-based plastics via self-exothermic effects: the origin and influential factors of exothermic processes. J. Hazard. Mater. 424, 127476.

Freeman, B.C., Bhown, A.S., 2011. Assessment of the technology readiness of post-combustion CO_2 capture technologies. Energy Procedia 4, 1791–1796.

Gu, F., Guo, J., Zhang, W., Summers, P.A., Hall, P., 2017. From waste plastics to industrial raw materials: a life cycle assessment of mechanical plastic recycling practice based on a real-world case study. Sci. Total Environ. 601-602, 1192–1207.

Henning, M., Haase, M., 2021. Techno-economic analysis of hydrogen enhanced methanol to gasoline process from biomass-derived synthesis gas. Fuel Process. Technol. 216, 106776.

International Energy Agency, 2021. Chemicals. https://www.iea.org/reports/chemicals, access December 2021.

Lin, Y., Jin, W., Cai, L., Liu, X., Qiu, Y., Zhang, G., 2020. Green preparation of covalently co-immobilized multienzymes on silica nanoparticles for clean production of reducing sugar from lignocellulosic biomass. J. Clean. Prod. 314, 127994.

Liu, W., Teng, L., Rohani, S., Qin, Z., Zhao, B., Xu, C.C., Ren, S., Liu, Q., Liang, B., 2021. CO_2 mineral carbonation using industrial solid wastes: a review of recent developments. Chem. Eng. J. 416, 129093.

Sanborn, A., 2019. Process for the Preparation of 2,5-Furandicarboxylic Acid (FDCA) by Oxidation of 5-(Alkyloxymethyl)Furfural (AMF) with Oxygen in the Presence of a

Co(II) or Ce(III) Catalyst, a Bromide Salt and a Solvent. Patent EP3133064B1, European Patent Office.

Tran, D.-T., Chang, J.-S., Lee, D.-J., 2017. Recent insights into continuous-flow biodiesel production via catalytic and non-catalytic transesterification processes. Appl. Energy 185, 376–409.

Vaz Jr., S., 2020. Treatment of Agroindustrial Biomass Residues—A Sustainable Approach. Springer Nature, Cham.

Yagushi, A.L., Lee, S.J., Blenner, M.A., 2021. Synthetic biology towards engineering microbial lignin biotransformation. Trends Biotechnol. 39, 1037–1064.

Yuan, R., Yu, S., Shen, Y., 2019. Pyrolysis and combustion kinetics of lignocellulosic biomass pellets with calcium-rich wastes from agro-forestry residues. Waste Manag. 87, 86–96.

CHAPTER 7

Carbon capture and use strategies for carbon dioxide

Technological strategies to mitigate the emissions of CO_2 and its effects on the environment—mainly, global warming—are of fundamental importance to decarbonize modern society and improve the quality of life and nature.

From this point of view, this chapter explore several strategies for various industry sectors to mitigate and create value chains.

7.1 Introduction to carbon capture and use strategies

According Vaz Jr. et al. (2022), carbon capture and use (CCU) strategies are aimed at CO_2 capture for industrial usage, with the technologies directed to capture this gas in the purest, economical, and environmentally viable way. Additionally, the stored gas can be used in carbon capture, use, and storage (CCUS). For the technology application, carbon dioxide capture can be carried out during three stages (Intergovernmental Panel on Climate Change, 2005):

- Precombustion
- Combustion
- Postcombustion

Precombustion is based on a gasification process through which the fuel passes and is intended to produce a synthesis gas (syngas), mainly composed of hydrogen (H_2) and carbon monoxide (CO). Subsequently, both gases are converted into CO_2, and then this goes through the gas separation process (Mohammad et al., 2020).

The capture of gases that occurs during combustion is called oxy-combustion, and its principle is the burning of fuel in an oxygen-enriched environment (Wilcox, 2012; Mohammad et al., 2020).

Postcombustion is the capture in the final phase of release of the combustion gases. It is ideal for capturing CO_2 from energy generation sources, such as thermoelectric plants and other plants that use fossil fuels and biomass to generate energy. After the combustion gases exit, they go through the

Renewable Carbon
https://doi.org/10.1016/B978-0-323-99735-5.00004-9

process of separating CO_2 from the other gases using the appropriate technology (Mohammad et al., 2020).

Defining the stage at which gas capture will occur is essential for the choice of the best carbon CCU/CCUS technologies, as shown in Fig. 7.1. Since these two factors are directly related, it is possible to provide the best technology that meets the criteria of cost, deployment, time, and abundance of CO_2 (or purity), promoting renewable carbon (RC) strategy.

It is worth mentioning that CCUS technologies can provide a method or process to supply CO_2 for conversion technologies as related to the catalytic routes, which will be discussed in Chapter 11 of Part 2.

The quantification—or a forecast for that—of the capacity of a certain technology to capture CO_2 should be stipulated to define its functionality (or capture feasibility) and utility (or quantity of CO_2 captured).

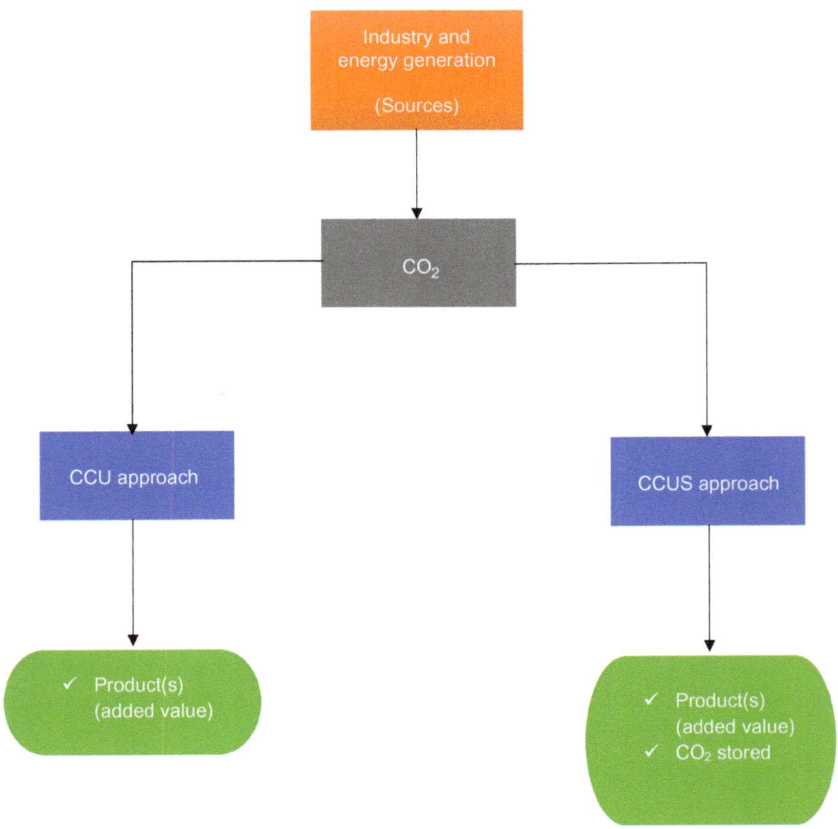

Fig. 7.1 Carbon capture systems for CO_2 gas reduction or mitigation to generate value. *CCU*, carbon capture and use; *CCUS*, carbon capture, use, and storage. *(Credit: author.)*

As a way to determine the quantity of carbon dioxide captured, Kemache et al. (2017) proposed Eqs. (7.1), (7.2).

$$CO_{2(\text{captured})}\ (\text{g}) = CO_{2(\text{inlet})}\ (\text{g}) - CO_{2(\text{output})}\ (\text{g}) \qquad (7.1)$$

$$CO_{2(\text{captured})}\ (\%) = CO_{2(\text{captured})}\ (\text{g})/CO_{2(\text{inlet})}\ (\text{g}) \times 100 \qquad (7.2)$$

The CO_2 inlet and outlet are measured and quantified according to those analytical technologies presented in Section 7.6 later in this chapter.

Regarding the global scenario related to the development and application of CCU and CCUS technologies, Fig. 7.2 depicts the behavior for one decade of observation.

In Fig. 7.2, we can observe an increase in operating status. As CCU/CCUS approaches involve generally high-cost technologies, the use of a technological maturity assessment is extremely important, as it identifies promising technologies capable of guiding investments in research, design, and implementation (RDI) (Roh et al., 2020) by analyzing the TRL levels the technology is capable of achieving (seen in the Chapter 6).

In this chapter, we will discuss the main categories of CCU/CCUS technologies to understand how to mitigate carbon dioxide emission from industrial sources, mainly those associated with energy production.

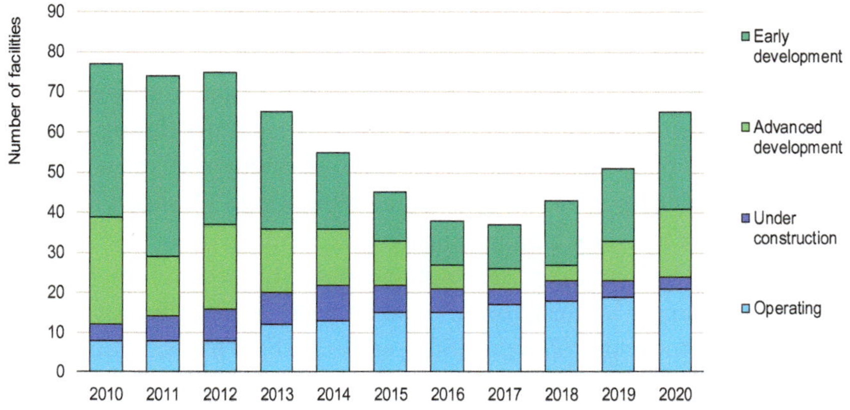

Fig. 7.2 World large-scale CCUS facilities operating and in development, 2010–20. *(Credit: International Energy Agency, 2021a. Carbon Capture, Utilization and Storage. https://www.iea.org/fuels-and-technologies/carbon-capture-utilisation-and-storage, Accessed December 2021. Reprinted with permission from the International Energy Agency.)*

7.2 Adsorption technologies

In adsorption, the gas comes into contact with porous particles capable of adsorbing it due to its surface affinity (e.g., similar chemical groups), separating carbon dioxide from the gas mixture. Table 7.1 shows examples of technologies based on the adsorption mechanism.

The adsorption processes are considered complex separation processes. Research efforts that aim to minimize the costs of these processes today present some limitations, among them the low efficiency in the gas separation process (Karimi et al., 2018). Another limiting factor in the development of adsorption technologies is the need for large amounts of energy to operate the process, that is, the energy balance is unfavorable.

Fig. 7.3 depicts a schematic diagram for the vacuum pressure swing adsorption (VPSA) technology.

7.3 Absorption technologies

In absorption, the process must occur in such a way that the carbon dioxide is capable of postcapture separation. This technology has materials capable of absorbing the gas and transporting it from one phase to another; in this case, the carbon dioxide is solubilized by passing to the liquid phase of the mixture, making it possible to separate it from the other components.

In thermoelectric plants, for example, the mixture of process gases is directed to a system in which the gases come into contact with amine solutions. This phase is intended to separate the CO_2 from the other components of the mixture by increasing the temperature (Silva et al., 2018).

Absorption is among the most promising technologies for its efficiency and economic viability. However, the energy expenditure during the process is also high (Silva and Santos, 2018). Because of this, research is aimed at optimizing the energy needed for the operation. Table 7.2 shows examples of technologies based on an absorption mechanism.

Absorption technologies are highly chosen by researchers for their advantages. These, in turn, occur mainly because absorption has been researched for more than 70 years (Silva et al., 2018). This condition makes absorption the most mature technology and capable of large-scale application (Freeman and Bhown, 2011).

Fig. 7.4 depicts the flow diagram for the application of absorption of CO_2 by monoethanolamine (MEA).

Table 7.1 Adsorption technologies.

Technology	Summary	Reference
In-situ capture and conversion of atmospheric CO_2 into nano-$CaCO_3$ using a novel pathway based on deep eutectic calcium chloride	In this study, CO_2 capture is carried out from a solvent solution based on eutectic choline chloride and has the purpose of making CO_2 a useful precursor for the production of calcite nanoparticles with added value. The results show feasibility for the precursor to be applied on a large scale, reducing CO_2 emissions and demonstrating the efficiency of the solution for the adsorption process.	Karimi et al. (2018)
Enhancement of CO_2 capture by using synthesized nano-zeolite	This research deals with the application of nano-zeolite in the process of CO_2 capture by adsorption, based on the temperature employed. Simulations were performed that varied parameters, such as percentage of adsorbent, CO_2 captured, and temperature. Keeping the CO_2 capture above 88% v/v, resulting in high efficiency of the nano-zeolite, in addition to the savings produced by the adsorbent.	Pham et al. (2016)
Development of carbon-based vacuum, temperature, and concentration swing adsorption postcombustion CO_2 capture processes	This work deals with vacuum, temperature, and concentration adsorption in a coal thermoelectric plant. Energy expenditure is considered the main impediment of this technology. To be able to evaluate the parameters used during the research, two scenarios were established, and the main result referred to the specific heat rate that presented itself as an advantage in relation to the use of conventional amines.	Plaza and Rubiera (2019)

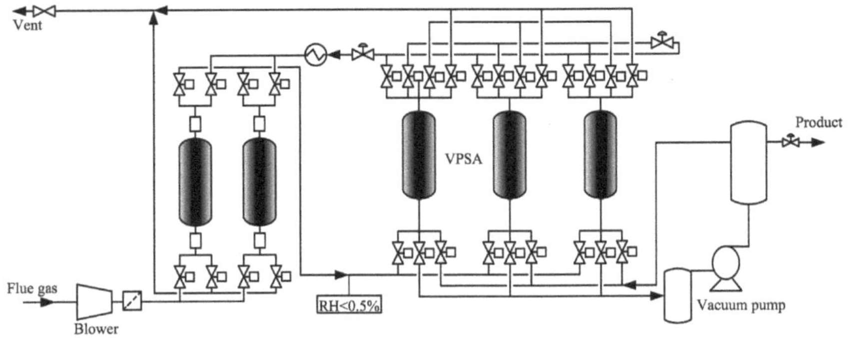

Fig. 7.3 Schematic diagram of pilot-scale carbon capture plant consisting of a dehumidifying unit and a CO_2 capture unit. The three-bed eight-step vacuum pressure swing adsorption (VPSA) process was employed in the CO_2 capture unit, where two types of adsorbent materials were tested, zeolite 13XAPG and 5A molecular sieve. *(Credit: Wang, L., Yang, Y., Shen, W., Kong, X., Li, P., Yu, J., Rodrigues, A.E., 2013. Experimental evaluation of adsorption technology for CO_2 capture from flue gas in an existing coal-fired power plant. Chem. Eng. Sci. 101, 615–619. Reprinted with permission from Elsevier.)*

7.4 Membrane separation technologies

The capture and separation of CO_2 by membranes is based on the selectivity of the material in relation to the gas. Carbon dioxide permeates through membranes, separating itself from other compounds. Due to their function, membranes depend on parameters such as medium permeability, particle size, selectivity, and pore volume (Nocito and Dibenedetto, 2019). Table 7.3 shows the main directions on these technologies, according to the literature.

The establishment of CO_2 capture by membranes is challenged by the development of systems that optimize the parameters necessary for the capture process. However, according to Wilcox (2012), this technology is considered the most effective among the existing ones.

Fig. 7.5 depicts a flow diagram for a membrane separation technology.

7.5 Chemical capture technologies

Among the carbon dioxide capture technologies currently under development, there is also the chemical capture for the generation of carbonates that can promote added value to industrial chains, in addition to leading to more sustainable production systems. This capture makes use of chemical reactions with carbon dioxide to obtain final products; these can be inorganic or organic, mainly carbonates. Table 7.4 shows technologies for the production of carbonates.

Table 7.2 Absorption technologies.

Technology	Summary	Reference
Results from a pilot plant using unpromoted potassium carbonate for carbon capture	This work aimed to study the capture of CO_2 by absorption using potassium carbonate as a solvent. The specificities of the process were analyzed from the incorporation of the chosen solvent. To trace the simulation parameters, the Aspen® software was used and percentages of the solvent were established to evaluate scenarios. The authors considered the results as promising to predict the performance of potassium carbonate in the absorption process.	Quyn et al. (2013)
Use of frothers to improve the absorption efficiency of dilute sodium carbonate slurry for postcombustion CO_2 capture	In this work, a pilot scale study was carried out for four types of solvent for the capture of CO_2 by absorption. During the tests, the authors incorporated a surfactant to evaluate it and also sparkling material. The capture efficiency in the scenario in question increased from 55% to 99%. The feasibility of this optimization was low cost, in addition to being environmentally appropriate.	Valluri and Kawatra (2020)
Experimental investigation on absorption performance of nanofluids for CO_2 capture	In this research, nanofluids were studied as CO_2 absorbers. The nanofluids used were those of TiO_2 and Al_2O_3. The main objectives of this research were to observe the types of nanoparticles and the influence of their concentrations on the absorption process. The result showed that increasing the concentration of these nanofluids decreased the relative CO_2 absorption index. To perform the comparison, a base fluid was used. Saline-based nanofluids were also tested, where it was found that this combination decreased the stability of the nanoparticles, reducing the efficiency of absorption.	Devakki and Thomas (2020)
CO_2 absorption rate and capacity of semiaqueous piperazine for CO_2 capture	This research studied piperazine, one of the most used solvents in CO_2 absorption technologies. Combinations with water and another solvent were carried out to evaluate the piperazine for its precipitation. The precipitation of this compound influences the efficiency of the gas absorption. The results showed that, due to the incorporation of new compounds, the CO_2 capture capacity and absorbent recycling capacity were increased.	Yuan and Rochelle (2019)

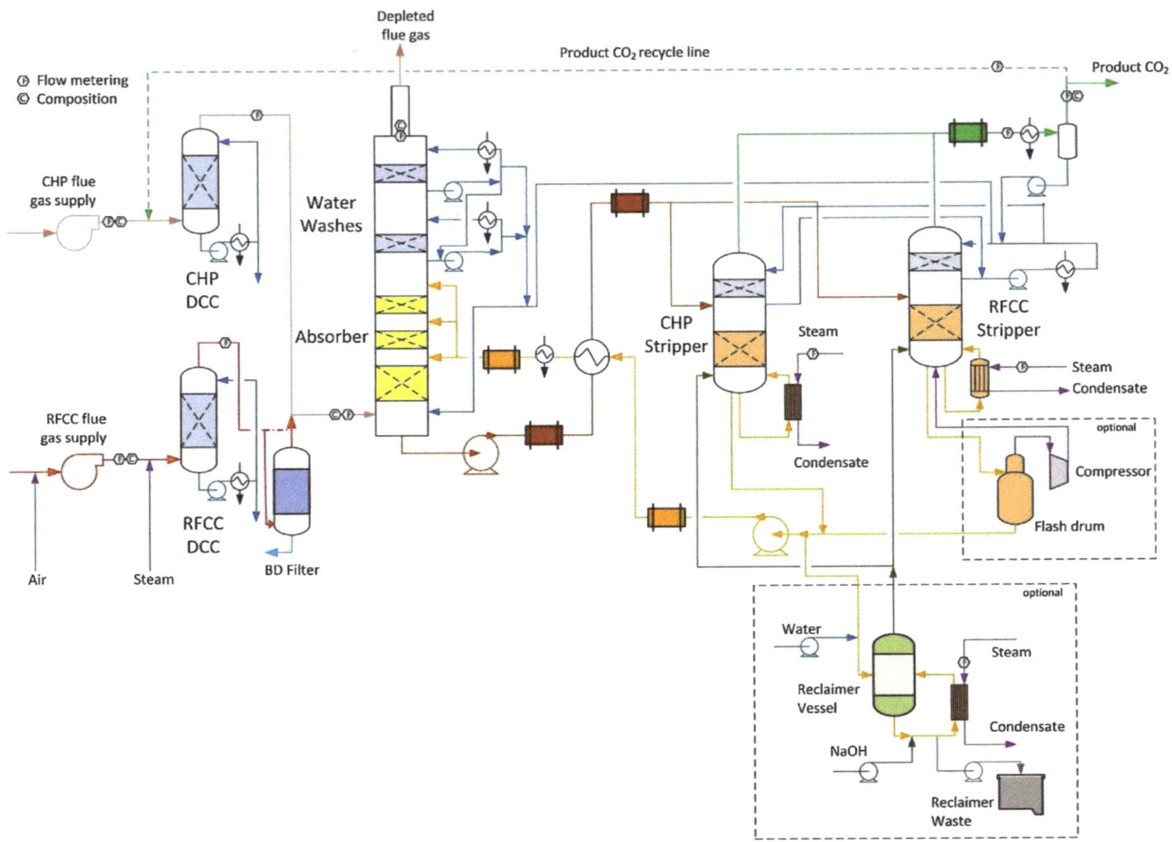

Fig. 7.4 Simplified process flow diagram of CO_2 capture with an aqueous monoethanolamine (MEA) solvent plant. *CHP*, combined heat and power; *DCC*, direct contact cooler; *RFCC*, residual fluidized catalytic cracker; *BD*, Brownian diffusion. *(Credit: Flø, N.E., Faramarzi, L., Iversen, F., Kleppe, E.R., Graver, B., Bryntesen, H.N., Johnsen, K., 2019. Assessment of material selection for the CO_2 absorption process with aqueous MEA solution based on results from corrosion monitoring at technology Centre Mongstad. Int. J. Greenhouse Gas Control 84, 91–110. Reprinted with permission from Elsevier.)*

Table 7.3 Membrane separation technologies.

Technology	Summary	Reference
Hybrid membrane process for postcombustion CO_2 capture from coal-fired power plant	This research is aimed at evaluating the selectivity of hybrid membranes that act in the process of capturing CO_2 and N_2 (which is also a greenhouse gas [GHG]) in thermoelectric plants. In this work, Aspen® software was used to simulate a thermoelectric plant, and the economic parameters of the technology, its permeability, and useful life were optimized. The result showed causality factors between the design of the plant and the properties of the membranes, in addition to identifying the selectivity of the membrane and cost variables.	Ren et al. (2020)
CO_2 capture by modified hollow fiber membrane contactor	In this work, the authors persuaded characteristics of membranes to be used to capture CO_2 from a natural gas source. A mathematical model was used to study elements such as membrane wettability, structure, type of absorbent used in the capture process, porosity, and fiber size. The simulations were carried out with four types of absorbents of varying fiber sizes; percentages were defined to estimate the most efficient values to define the ideal membrane. The results presented showed that absorbent concentration, density, and porosity directly influenced the separation of CO_2.	Abdolahi-Mansoorkhani and Seddighi (2020)
Mass transfer characteristics of a continuously operated hollow-fiber membrane contactor and stripper unit for CO_2 capture	The focus of this work was to evaluate the mass transfer of a hollow fiber membrane contactor. The parameters that directly influenced the mass transfer coefficient were explored to make the efficiency of the contactor visible when connected to the membrane. From the tests, it was possible to observe the workability of the contactor used. It had many limitations regarding desorption. Therefore, it was suggested that vacuum tests were inefficient, requiring the moistening of the membranes to facilitate the transfer of mass through them.	Nieminen et al. (2020)

Continued

Table 7.3 Membrane separation technologies—cont'd

Technology	Summary	Reference
Membrane gas-solvent contactor pilot plant trials for postcombustion CO_2 capture	This research involved an absorption test carried out in a pilot plant for the generation of energy from mineral coal. Here, the authors used gas-solvent contactors to promote greater efficiency and compactness in the CO_2 capture process through hybrid membrane capture technology. The contactors were able to assist the solvents in the gas absorption process. As a result, the contactors proved to be efficient when applied on an industrial scale.	Scholes et al. (2020)

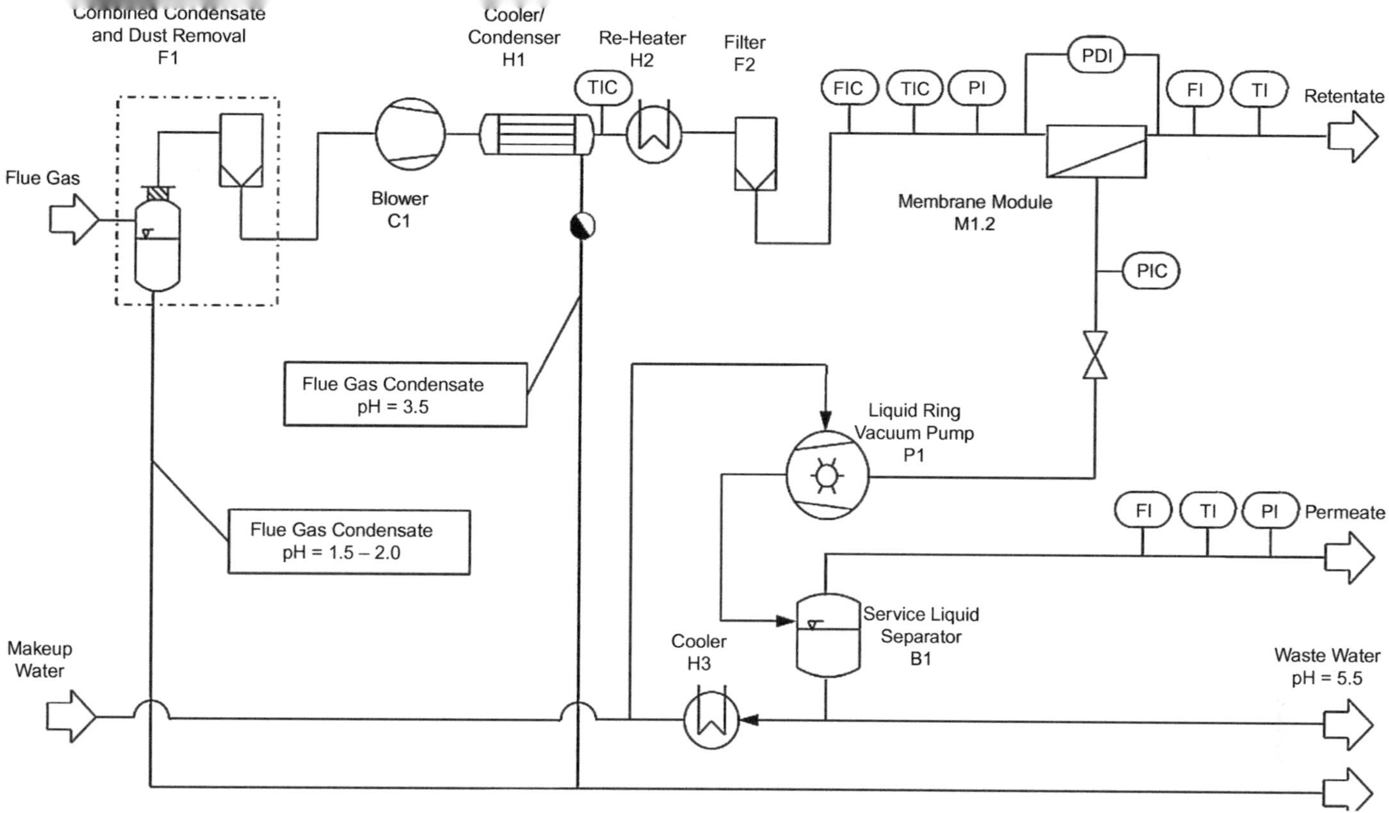

Fig. 7.5 Simplified flow sheet of the pilot plant for membrane separation technology. The membrane used in the experiments was a multilayer thin film composite membrane with an active separation layer made from PolyActive™. The support structure consists of a nonwoven support with a thickness of app. 100 μm made from polyester (PES), a porous support layer of polyacrylonitrile (PAN) with a thickness of approx. 50 μm, and an average pore size of 30 nm on the membrane surface. *FIC*, flowrate controller; *TIC*, temperature controller; *PIC*, pressure transducer. *(Credit: Pohlmann, J., Bran, M., Wilkner, M., Brinkmann, T., 2016. Pilot scale separation of CO_2 from power plant flue gases by membrane technology. Int. J. Greenhouse Gas Control 53, 56–64. Reprinted with permission from Elsevier.)*

Table 7.4 Chemical capture technologies.

Technology	Summary	Reference
Experimental investigation of the carbonation reactor in a tail-end calcium looping configuration for CO_2 capture from cement plants	In this work, CO_2 capture through calcium looping technology was investigated. The research was conducted on a semi-industrial scale in thermoelectric plants burning fossil fuels. The results showed that higher concentrations of sorbent increase the transport of gas. The tests validated the technology used to capture CO_2, with the same being shown for application in cement manufacturing plants.	Hornberger et al. (2020)
CO_2 capture using Ca-looping with CaO and Al_2O_3	To optimize the Ca-looping technology, this work investigated the use of additives to reduce the deactivation of CaO during the CO_2 capture process. The results obtained showed that the combinations of carbonates with Al_2O_3 additive used in this experiment were effective in capturing CO_2, since it prevented the closure of pores and the growth of CaO crystals.	Silva and Santos (2018)
Carbon dioxide capture using calcium oxides	This work aimed to evaluate the incorporation of calcium oxide to the CO_2 capture process by carbonation. Two carbonation combinations were made: one with pure calcium and the other with calcium mixed with alumina. It was found that carbonation with pure calcium promoted greater CO_2 capture than the mixture with alumina, which proved to be inefficient in the process of capturing CO_2 as carbonate.	Bisinoti et al. (2017)
Single-step, low temperature, and integrated CO_2 capture and conversion using sodium glycinate to produce calcium carbonate	This work studied the capture and conversion of CO_2 to produce calcium carbonate with CaO and $CaSiO_3$ as precursors. Calcium glycinate was used to promote CO_2 capture, and parameters such as temperature, concentration of the additive, and reaction time were investigated. The results obtained suggested that glycinate undergoes multiple cycles of capture and regeneration, favoring the precipitation of carbonates.	Liu and Gadikota (2020)

The great interest in this class of technology is because the carbonates obtained have a high boiling point, low toxicity, and are biodegradable (North et al., 2010).

Fig. 7.6 depicts a block diagram for the chemical-looping (CL) technology. CL uses an oxygen transfer material—also known as oxygen carrier—that drives the reactions in a cyclic process to supply the heat necessary by the oxidation reactions feeding air to the reactor, by avoiding the use of external burners to heat up the reformer, and the costs associated with the use of highly corrosion-resistant materials (Palma et al., 2020).

7.6 Analytical technologies for CO_2 monitoring

The process analytical technology (PAT) is the best approach for monitoring CO_2 in CCU/CCUS processes. It is possible to measure CO_2 by means an analytical technology with the Raman spectroscopy coupled to the respective reactor in an online mode (Fig. 7.7).

As observed by Jinadasa et al. (2018), the implementation of PAT for CCU/CCUS technologies, e.g., absorption, requires success in five elements. They are as follows:

- Selecting an appropriate process analyzer
- Integration between the analyzer and the process

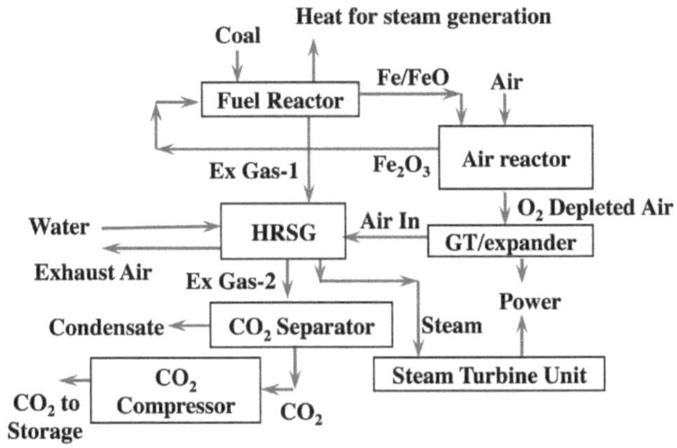

Fig. 7.6 Simplified block diagram for the coal direct-chemical looping combustion (CDCLC) process. *HRSG*, heat recovery steam generator; *GT*, gas turbine. *(Credit: Mukherjee, S., Kumar, P., Yang, A., Fennell, P., 2015. Energy and exergy analysis of chemical looping combustion technology and comparison with pre-combustion and oxy-fuel combustion technologies for CO_2 capture. J. Environ. Chem. Eng. 3, 2104–2114. Reprinted with permission from Elsevier.)*

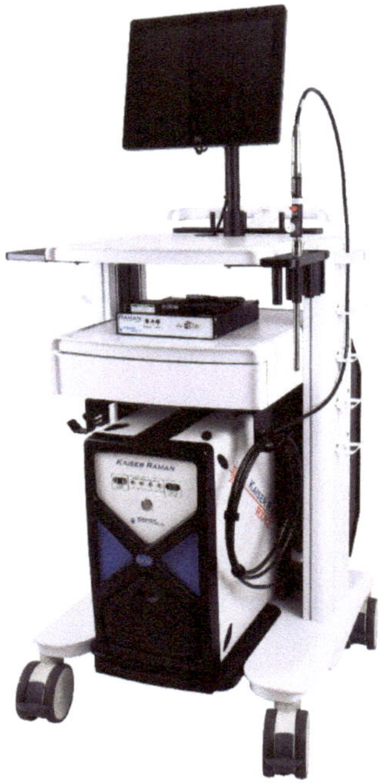

Fig. 7.7 A Kaiser Raman Rxn2 analyzer for PAT application to CCU/CCUS processes. *(Courtesy of Kaiser Optical Systems, Inc.)*

- Model development to enable the analyzer to predict a process-related chemical or physical attribute
- Use of the developed model in real-time application
- Use of the data obtained from the analyzer as an input to a process control unit

Furthermore, partial least square regression model is a useful chemometric-based method to extract hidden chemical information in measurements from a process analyzer.

More information about chemometrics methods for spectroscopy can be obtained in the book "Chemometrics in Spectroscopy" (authored by Mark and Workman, Jr., and published by Elsevier).[a]

[a] https://www.elsevier.com/books/chemometrics-in-spectroscopy/mark/978-0-12-805309-6.

7.7 Case study: The TRL assessment for CCU technologies for energetic sources

The International Energy Agency (2020) carried out a survey of perspectives on CO_2 capture technologies according to TRL levels. From this survey, the TRL assessment for the four categories of technologies presented here (adsorption, absorption, membrane separation, and chemical capture) is presented in Table 7.5.

TRL surveys were conducted using data from technology providers. For a critical evaluation purpose, statements from Kearns et al. (2021) were considered to clarify that all elements of the carbon capture and storage value chain are mature and have been in commercial operation for decades. However, incremental improvements in those technologies have, and will continue, to reduce their cost.

As previously observed in the Chapter 6, the use of tools like the TRL scale changes the perspective of the RDI efforts and investments. Advantages such as time and cost projections stand out in their application and are extremely valid in monitoring the development of GHG capture and use technologies.

Table 7.5 TRL assessment for CCU/CCUS technologies based on the technological survey carried out by the International Energy Agency (2020).

Technology category	TRL level	Considerations
Adsorption	9	Mainly applied in natural gas and ethanol processes, this technology is responsible for CO_2 capture in large plants and has great application perspectives. Its advances are mainly due to the simple operation attributed to it.
Absorption	9	It is the most advanced technology. This is due to the research time and consequently its application in small and large power generation, fuel transformation, and industrial production plants.
Membrane separation	6–7	Relatively new but promising technology and considered to be the most effective separation technology among the existing ones. Its advances depend on the type of gas emission source and its application. Currently, part of its applications are in the demonstration phase and another part in the development phase; few are commercially available.
Chemical capture	4–6	The capture involving chemical reactions are presented in that TRL for its time and research intensity. As it is relatively new, its level is justified by the need for large-scale pilot tests.

Reprinted with permission from the International Energy Agency.

Furthermore, net-zero energy system by 2050 should ensure stable and affordable energy supplies, providing universal energy access, and enabling robust economic growth (International Energy Agency, 2021). To reach these goals, CCU/CCUS technologies have a relevant role because systems of production will continue emitting GHG—not limited to CO_2—despite the global efforts to reduce their emissions.

Fig. 7.8 depicts the capacity and storage times for several sequestration methodologies; some of them were not considered in this book because they

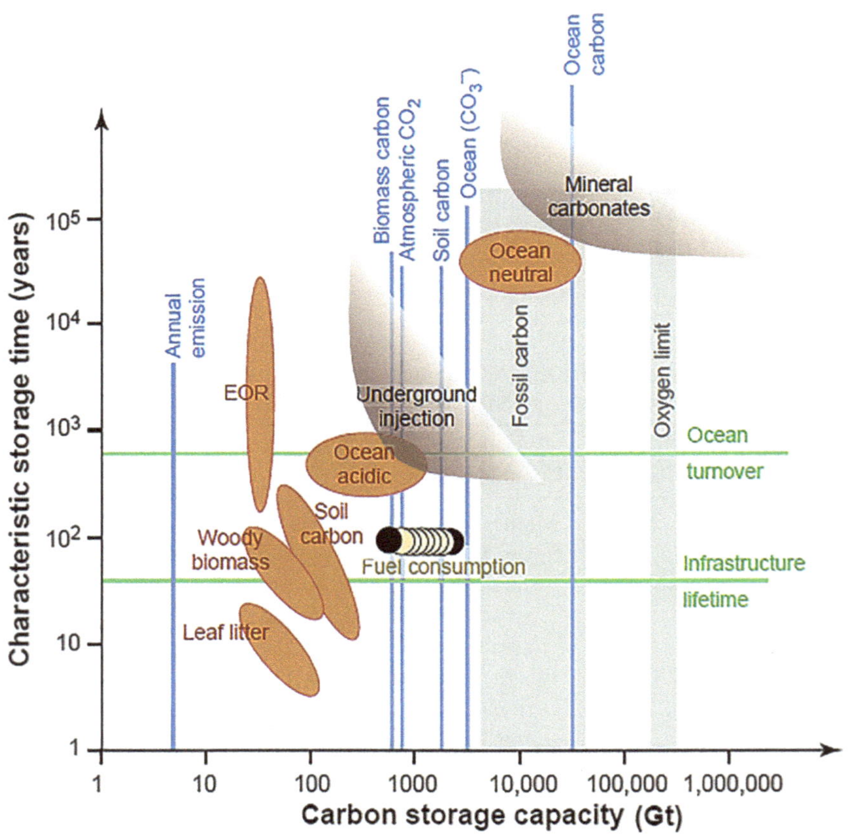

Fig. 7.8 Estimated storage capacities and times for various sequestration methods. The "fossil carbon" range includes methane hydrates from the ocean floor at its upper end. The "oxygen limit" is the amount of fossil carbon that would use up all oxygen available in air for its combustion. Carbon consumption for the 21st century ranges from 600 Gt (current consumption held constant) to 2400 Gt. "Ocean acidic" and "ocean neutral" are the ocean's uptake capacities for carbonic acid and neutralized carbonic acid, respectively. The upper limits of capacity or lifetime for underground injection and mineral carbonates are not well constrained. EOR stands for enhanced oil recovery. *(Credit: Lackner, K.S., 2003. A guide to CO_2 sequestration. Science 300, 1677–1678. Reprinted with permission from American Association for the Advancement of Science.)*

are out of its scope. From this content, we can observe the urgency to put into practice technologies of CCU/CCUS, which permit capture and use in a reduced time to attend the global demand to reduce GHG emission, mainly for CO_2.

7.8 Conclusions

Technologies for CCU/CCUS are paramount for industrial and energy sectors to permit CO_2 mitigation and its use as RC raw material.

Despite its importance, CCU/CCUS approaches comprise technology categories with limitations and challenges for their application. However, we can observe an increase in their use and the development of more sustainable and affordable technologies.

References

Abdolahi-Mansoorkhani, H., Seddighi, S., 2020. CO_2 capture by modified hollow fiber membrane contactor: numerical study on membrane structure and membrane wettability. Fuel Process. Technol. 209, 106530.

Bisinoti, L.V., Silva, J.A., Santos, J.B.O., 2017. Captura de dióxido de carbono utilizando óxidos a base de cálcio [Capture of carbon dioxide using calcium oxides]. In: XII Congresso Brasileiro de Engenharia Química em Iniciação Científica. UFSCar, São Carlos, SP, pp. 2902–2907.

Devakki, B., Thomas, S., 2020. Experimental investigation on absorption performance of nanofluids for CO_2 capture. Int. J. Air-Cond. Refrig. 28, 2050017.

Freeman, B.C., Bhown, A.S., 2011. Assessment of the technology readiness of post-combustion CO_2 capture technologies. Energy Procedia 4, 1791–1796.

Hornberger, M., Moreno, J., Schmid, M., Scheffknecht, G., 2020. Experimental investigation of the carbonation reactor in a tail-end calcium looping configuration for CO_2 capture from cement plants. Fuel Process. Technol. 210, 106557.

Intergovernmental panel on climate change, 2005. IPCC Special Report on Carbon Dioxide Capture and Storage. Prepared by Working Group III of the Intergovernmental Panel on Climate Change. Cambridge University Press, Cambridge. 442 p.

International Energy Agency, 2020. Energy Technology Perspectives 2020. https://www.iea.org/reports/energy-technology-perspectives-2020. accessed December 2021.

International Energy Agency, 2021. Net Zero by 2050: A Roadmap for the Global Energy Sector. https://www.iea.org/reports/net-zero-by-2050. accessed December 2021.

Jinadasa, M.H.W.N., Jens, K.-J., Halstensen, M., 2018. Process Analytical Technology for CO_2 Capture. https://www.intechopen.com/chapters/60759. accessed December 2021.

Karimi, M., Jodaei, A., Khajvandi, A., Sadeghinikrahim, A., Jahandideh, R., 2018. In-situ capture and conversion of atmospheric CO_2 into nano-$CaCO_3$ using a novel pathway based on deep eutectic choline chloride-calcium chloride. J. Environ. Manage. 206, 516–e522.

Kearns, D., Liu, H., Comsoli, C., 2021. Technology Readiness and Costs of CCS. Global CCS Institute. https://www.globalccsinstitute.com/wp-content/uploads/2021/03/Technology-Readiness-and-Costs-for-CCS-2021-1.pdf. (accessed December 2021).

Kemache, N., Pasquier, L.-C., Cecchi, E., Mouedhen, I., Blais, J.-F., Mercier, G., 2017. Aqueous mineral carbonation for CO_2 sequestration: from laboratory to pilot scale. Fuel Process. Technol. 166, 209–216.

Liu, M., Gadikota, G., 2020. Single-step, low temperature and integrated CO_2 capture and conversion using sodium glycinate to produce calcium carbonate. Fuel 275, 17887.

Mohammad, M., Isaifan, R.J., Weldu, Y.W., Rahman, M.A., Al-Ghamdi, S.G., 2020. Progress on carbon dioxide capture, storage and utilisation. Int. J. Glob. Warm. 20 (2), 124.

Nieminen, H., Järvinen, L., Ruuskanen, V., Laari, A., Koiranen, T., Ahola, J., 2020. Mass transfer characteristics of a continuously operated hollow-fiber membrane contactor and stripper unit for CO_2 capture. Int. J. Greenhouse Gas Control 98, 103063.

Nocito, F., Dibenedetto, A., 2019. Atmospheric CO_2 mitigation technologies: carbon capture utilization and storage (CCUS). Curr. Opin. Green Sustain. Chem. 21, 34–43.

North, M., Pasquale, R., Young, C., 2010. Synthesis of cyclic carbonates from epoxides and CO_2. Green Chem. 12, 1514–1539.

Palma, V., Barba, D., Meloni, E., Renda, S., Ruocco, C., 2020. Chapter 17—Ultracompact biofuels catalytic reforming processes for distributed renewable hydrogen production. Stud. Surf. Sci. Catal. 179, 317–333.

Pham, T.-H., Lee, B.-K., Kim, J., Lee, C.-H., 2016. Enhancement of CO_2 capture by using synthesized nano-zeolite. J. Taiwan Inst. Chem. Eng. 64, 220–226.

Plaza, M.G., Rubiera, F., 2019. Development of carbon-based vacuum, temperature and concentration swing adsorption post-combustion CO_2 capture processes. Chem. Eng. J. 375, 122002.

Quyn, D., Rayer, A.V., Gouw, J., Indrawn, I., Mumford, K.A., Anderson, C.J., Hooper, B., Stevens, G.W., 2013. Results from a pilot plant using un-promoted potassium carbonate for carbon capture. Energy Procedia 37, 448–454.

Ren, L.-X., Chang, F.-L., Kang, D.-Y., Chen, C.-G., 2020. Hybrid membrane process for post-combustion CO_2 capture from coal-fired power plant. J. Membr. Sci. 603, 118001.

Roh, K., Bardoe, A., Bongartz, D., Burre, J., Chung, W., Deutz, S., Han, D., Hebelmann, M., Kohlhaas, Y., Konig, A., Lee, J.S., Volker, S., Wessling, M., Lee, J.H., Mitsos, A., 2020. Early-stage evaluation of emerging CO_2 utilization technologies at low technology readiness levels. Green Chem. 22, 3842–3859.

Scholes, C.A., Kentish, S.E., Qader, A., 2020. Membrane gas-solvent contactor pilot plant trials for post combustion CO_2 capture. Sep. Purif. Technol. 237, 116470.

Silva, J.A., Santos, J.B.O., 2018. Captura de CO_2 Utilizando O Processo ca-Looping Com CaO e Al_2O_3 [CO_2 Capture Using Ca-Looping with CaO and Al_2O_3]. Universidad Nacional de Cuyo. XXVI Jornada de Jóvens investigadores, Argentina.

Silva, W.W.M., Barros, G.N., Pinto, J.R.S., Camargos-Filho, J.F., Rezende, K.I., Menezes, M.T.S., Silqueira, M.C.S., Nunes, V.H.S., Dias, J.F., Mogueira, E., Rabelo, S.N., 2018. Análise e levantamento das principais técnicas para captura do CO2 [Analysis and survey of the main techniques for CO2 capture]. In: Semana de Iniciação Científica e Tecnológica. Universidade de Itaúna, Brazil.

Valluri, S., Kawatra, S.K., 2020. Use of frothers to improve the absorption efficiency of dilute sodium carbonate slurry for post combustion CO_2 capture. Fuel Process. Technol. 212, 106620.

Vaz Jr. S., de Souza, A.P., Baeta, BEL., 2022. Technologies for carbon dioxide capture: A review applied to energy sectors. Clean. Eng. Technol. 8, 100456.

Wilcox, J., 2012. Carbon Capture. Springer Nature, Cham.

Yuan, Y., Rochelle, G.T., 2019. CO_2 absorption rate and capacity of semi-aqueous piperazine for CO_2 capture. Int. J. Greenhouse Gas Control 85, 182–186.

CHAPTER 8

Applied nanotechnology for carbon dioxide, biomass, and recycling materials

Nanotechnology is one of the most promising technological branches for the 21st century, reaching several industrial sectors and ending uses. The application of nanotechnology to renewable carbon (RC) raw materials is also expected to be a strong opportunity for new value chains, as demonstrated in this chapter.

8.1 Introduction to nanoscience and nanotechnology

Nanoscience (the branch definition) and nanotechnology (the branch application) are subjects of interest from academia and industry due to their potential for the development of new materials as well as the potential to generate new value chains, such as those derived from RC raw materials. For example, the development of catalysts as functionalized nanoparticles (NPs) obtained from biomass is extremely strategic to convert the biomass into valuable products and meet the green chemistry principles (Shuttleworth et al., 2014).

To reinforce an academic definition, nanoscience can be understood as the study of matter's properties at the nanoscale—generally considered within 1–100 nm; in particular, it focuses on the unique, size-dependent properties of solid-state materials (Mulvaney, 2015), with improved functionalization when compared to microscale and superior scales of molecular size.

Additionally, according the National Institute for Occupational Safety and Health (2021), nanotechnology can be defined as the manipulation of matter on a near-atomic scale to produce new structures, materials, and devices applied to many sectors, such as:

- Medicine: developing new methods of preventing, diagnosing, and treating various diseases (Zdrojewicz et al., 2015)

Renewable Carbon
https://doi.org/10.1016/B978-0-323-99735-5.00010-4

- Consumer products: goods for children, appliances, food and beverage, electronics, cross-cutting, automotive, home and garden, health and fitness (supplements, sunscreens, filtration, sporting goods, cosmetics, clothing, personal care) (Vance et al., 2015)
- Energy: battery efficiency, efficient combustion and reduced emissions, plastic solar cells, and fuel cells (National Nanotechnology Initiative, 2021a)
- Materials: biomaterials, ceramic powders, corrosion and protection, electroceramics, high purity rare earths, magnetic and electric materials, and polymeric materials (Institute for Energy and Nuclear Research, 2021)
- Manufacturing: nanomanufacturing involving scaled-up, reliable, and cost-effective manufacturing of nanoscale materials, structures, devices, and systems according to top-down processes and increasingly complex bottom-up or self-assembly processes (National Nanotechnology Initiative, 2021b)

Nanotechnology is generally understood as engineered structures, devices, and systems, whereas nanomaterials are defined as those things that have, as previously described, a length scale between 1 and 100 nm. At this size, materials begin to exhibit unique properties that affect physical, chemical, and biological behavior.

Fig. 8.1 illustrates the nanoscale of several common items.

Particle dimensions in the range of 1–100 nm (International Standard Organization, 2008) can confer some interesting molecular properties to be explored in several applications, such as (Hassellöv and Kaegi, 2009):

- Surface area
- Porosity
- Structure
- Crystallinity
- Size distribution
- Surface charge
- Surface speciation

It is expected that a technology readiness level (TRL) ranges from TRL3 (development) to TRL6 (demonstration) for most of the nanotechnologies observed in the scientific and patent literature, and industry statements.

Fig. 8.2 depicts an image obtained from a nanocarbonate produced from CO_2 chemical conversion, as an example of CCU strategy (see Chapter 7) in nanoscale of application.

However, aspects of toxicology and ecotoxicology of the NPs are a subject of intense discussion, and understanding and use of NPs production and application will likely soon increase. To subsidize the debate and guide the

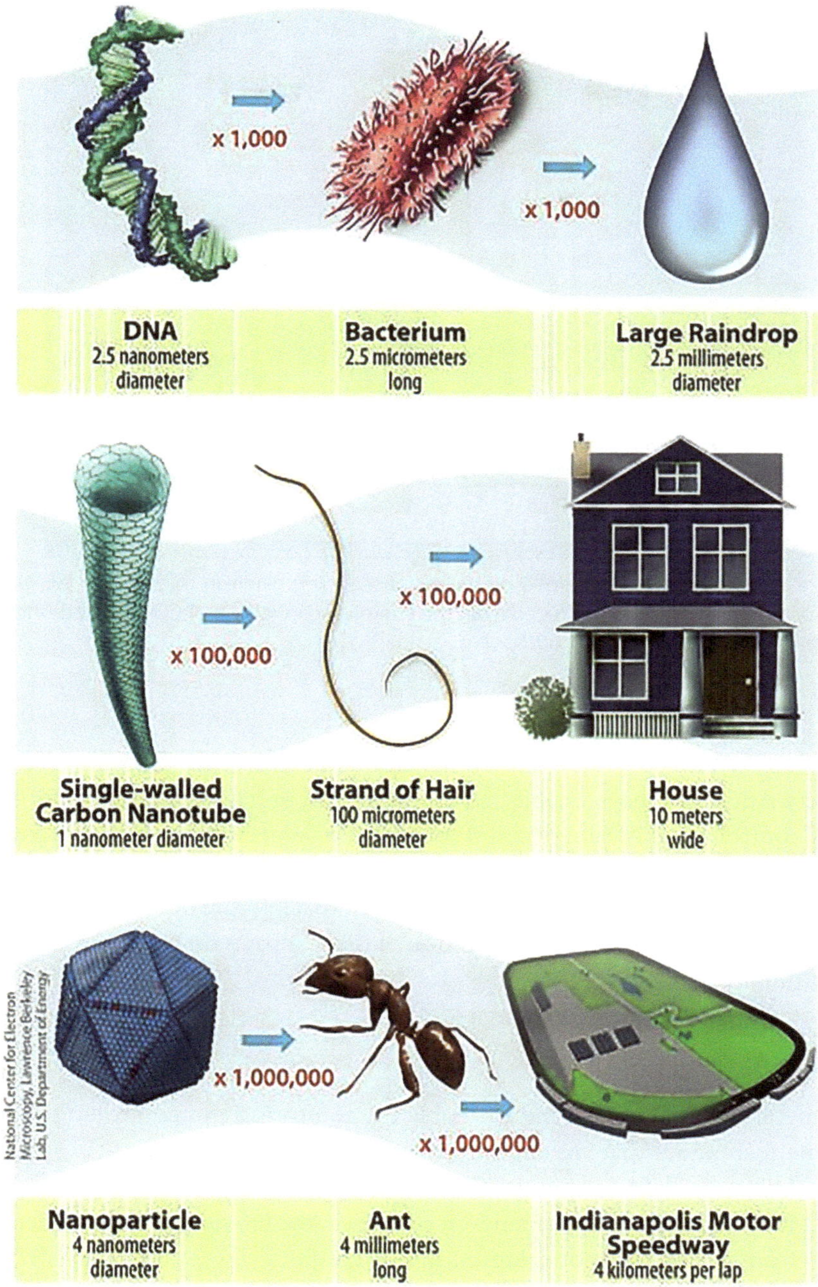

Fig. 8.1 Visual examples of the size and the scale of nanomaterials. *(Credit: U.S. Department of Energy. Reprinted with permission from the U.S. Government.)*

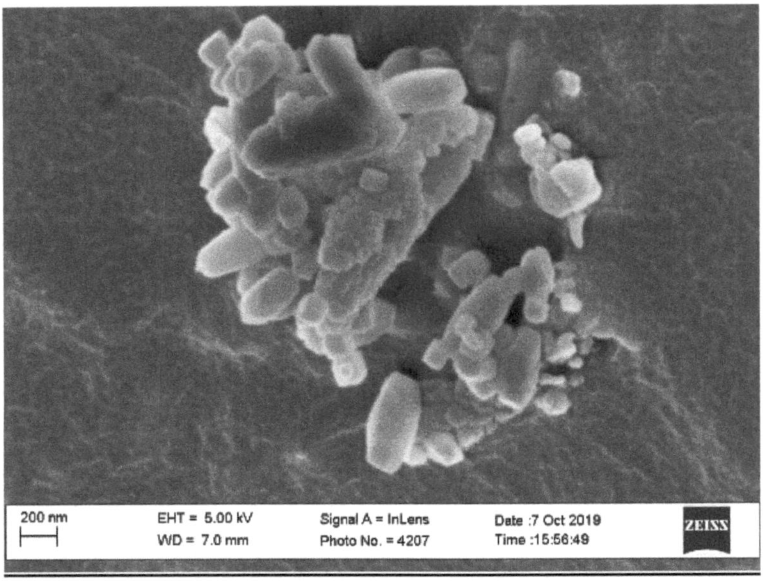

Fig. 8.2 A micrography obtained by means of scanning electron microscopy (SEM) of nanocarbonate particles produced from the chemical conversion of CO_2 emitted from a thermoelectric plant. *(Credit: author.)*

productive sectors, some international agencies, such as the European Food Safety Authority, have established requisites and procedures for monitoring and controlling this category of new products, especially for feed and food (Quik et al., 2020).

The risk assessment for NPs is paramount to allow the best usage of these materials, by means of a clear definition of their impacts on the environment and public health.

Analytical techniques

Analytical techniques are indispensable to monitoring and controlling NPs. Some techniques that are commonly used are:

- Dynamic light scattering (DLS): for determining the particle size by means of measuring the random changes in the intensity of light scattered from a suspension or solution; this technique is also called photon correlation spectroscopy (PCS) and quasi-elastic light scattering (QELS).
- Thermogravimetric analysis (TGA): to observe the thermal decomposition of the sample by means of the mass loss according to the temperature increase.

- Surface area and volume of pores analysis: for determining the physicochemical properties with the interpretation of the data using the Brunauer-Emmett-Teller (BET) theory of gas adsorption in multilayers (Brunauer et al., 1983).
- Differential scanning calorimetry (DSC): for determining the glass transition temperature of the sample—a physicochemical property; the difference in the amount of heat required to increase the temperature of a sample and a reference material is measured as a function of temperature.
- Size exclusion liquid phase chromatography with light scattering detector (SEC-ELSD): for observing molecule size distribution—mainly for polymeric material—to determine the physicochemical property.
- Fourier transformed infrared absorption spectroscopy (FTIR): for structural elucidation by means of infrared absorption producing absorption bands based on vibrational modes characteristic for certain chemical groups according to their polarity.
- Scanning electron microscopy (SEM) combined with the identification of chemical elements present in the sample surface by energy-dispersive X-ray spectroscopy (EDS): for observing the morphology of the sample surface to determine the physicochemical property and its chemical composition.
- Transmission electron microscopy (TEM): as with SEM, it is a microscopy technique that permits higher magnifications than SEM and also shows the crystallographic structure and composition of a material.

The analytical approach and, consequently, the best techniques for use will depend on the sample's characteristics (e.g., physical state). Eventually, other techniques not listed here can be used. *Handbook of Instrumental Techniques for Analytical Chemistry* (edited by Frank Settle and published by Prentice Hall)[a] can be very useful to choose the most suitable technique(s).

8.2 Nanotechnology applied to CO_2: A case study

This case study is based on the Patent WO2013/090274A1 from ExxonMobil (Denton and Noyes, 2013) related to a method for forming carbon nanotubes (Fig. 8.3). The method includes forming carbon nanotubes in a first reactor by using a feed gas (CO_2 and CH_4). The carbon nanotubes are separated from the reactor effluent to form a waste stream. The feed gas, a dry

[a] https://www.tandfonline.com/doi/abs/10.1080/10826079808006889?journalCode=ljlc20.

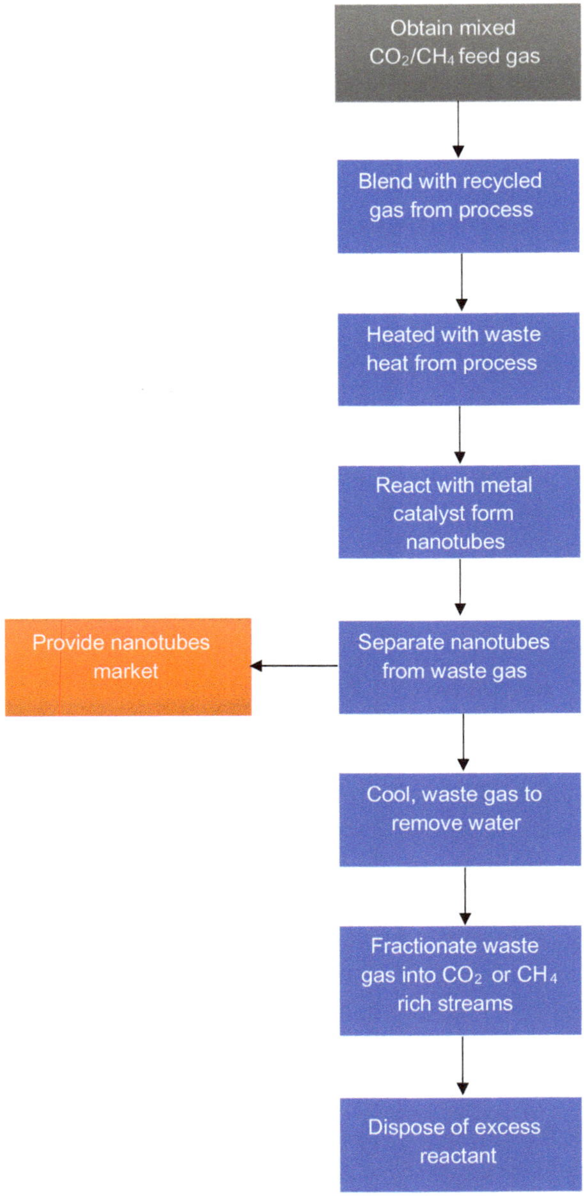

Fig. 8.3 Flow diagram of the ExxonMobil's method to obtain carbon nanotubes from a mixture of CO_2/CH_4. *(Reprinted from Denton, R.D., Noyes, D., 2013. Method and System for Forming Carbon Nanotubes. Patent WO2013/090274A1. Patentscope. World Intellectual Property Organization, with permission from the World Intellectual Property Organization.)*

waste stream, or both are heated with waste heat from the waste stream. The waste stream is chilled in an ambient temperature heat exchanger to condense water vapor, forming a dry waste stream.

According to Abdalla et al. (2015), properties of nanostructured carbon nanotubes (Fig. 8.4) such as high surface area, good stiffness, and resilience have been explored in many engineering applications. Research on carbon nanotubes has shown applications in the fields of energy storage, hydrogen storage, electrochemical supercapacitor, field-emitting devices, transistors, nanoprobes and sensors, composite material, and templates, among others.

Then this type of material—introduced in Chapter 3—has considerable technological and economic potentials; its association with CO_2 mitigation, as proposed by ExxonMobil in its patent, can promote a sustainable material for the global market. Moreover, the use of CO_2 combined with CH_3 can be a suitable approach, for example, to promote the mitigation of greenhouse gases (GHG) in petrochemical plants by means of capturing and using both gases.

8.3 Nanotechnology applied to biomass derivatives: A case study

Lignin, the second most abundant biomass component and a phenolic macromolecule, can be converted into a functionalized, lignin-derived, carbon-based adsorbent by means of hydrothermal carbonization using

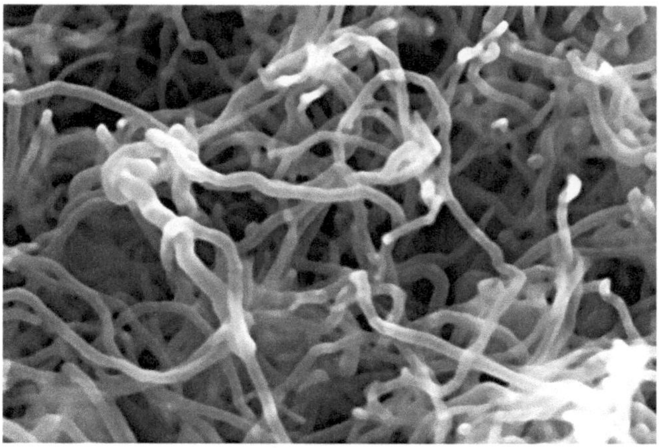

Fig. 8.4 Carbon nanotube bundles obtained by scanning electron microscopy (SEM). *(Credit: Shutterstock, https://www.shutterstock.com/pt/image-photo/nanotubes-bundles-real-photo-electron-microscope-86827882.)*

MgO NPs. And this material can be used to remove phosphorus (as PO_4^{3-}) from industrial aqueous effluent, as proposed by Jiao et al. (2021). The phosphate efficiency removal of 99.76% was achieved. Fig. 8.5 depicts the system and its application.

This type of material is very useful for environmental applications to control pollutants in water and wastewater. In the case of PO_4^{3-} ion, it can promote water eutrophication with implications to the environment and health because, with the increasing availability of nutrients, we have a considerable increase in the number of algae and cyanobacteria, which

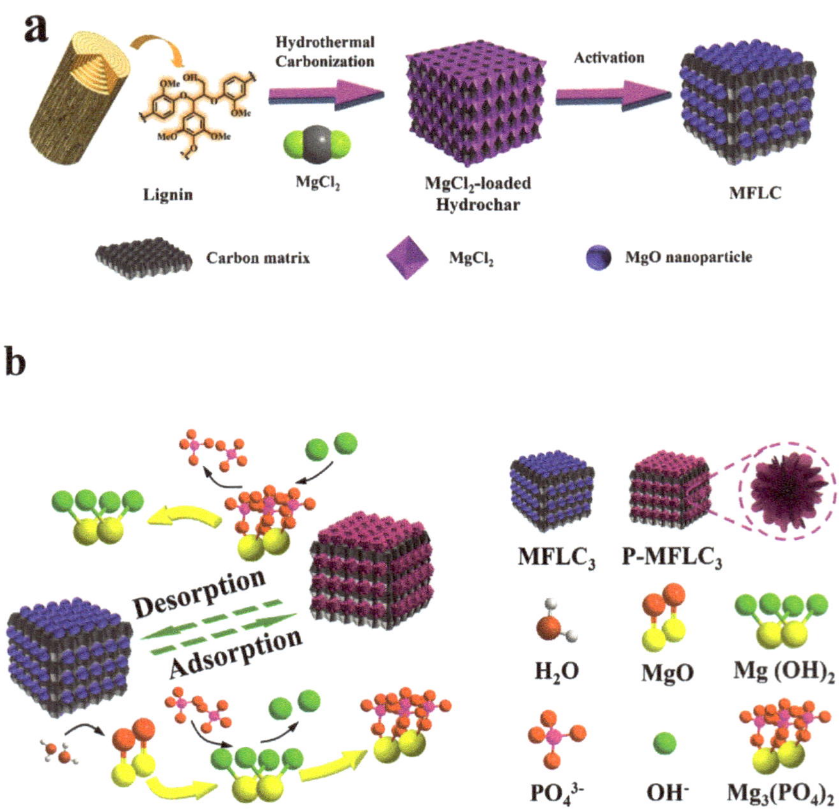

Fig. 8.5 The lignin-derived, carbon-based adsorbent. (A) Adsorbent production; (B) adsorbent activity. $MFLC_n$ = functionalized lignin-derived carbon-based charcoal; P-$MFLC_n$ = phosphorus-functionalized lignin-derived carbon-based. *(Reprinted from Jiao, G.-J., Ma, J., Li, Y., Jin, D., Guo, Y., Zhao, J., Sun, R., 2021. Enhanced adsorption activity for phosphate removal by functional lignin-derived carbon-based adsorbent: optimization, performance and evaluation. Sci. Total Environ. 761, 143217, with permission from Elsevier.)*

are phosphorus participates in fundamental processes in the metabolism of living beings, such as energy storage and cell membrane formation.

8.4 Nanotechnology applied to recycling materials: A case study

This case study is based on research conducted by Algozeeb et al. (2020) related to the method that uses flash Joule heating (FJH) to convert plastic waste (PW) into flash graphene (FG) (Fig. 8.6). In addition to FG, the process results in the formation of carbon oligomers, hydrogen, and light hydrocarbons.

To make high-quality graphene, a sequential alternating current (AC) and direct current (DC) flash was used. The FJH process requires no catalyst and performs on PW mixtures, which makes the process suitable for handling landfill PW. The energy required to convert PW to FG is $\sim 23\,\text{kJ}\,\text{g}^{-1}$ or US \$125 in electricity per ton of PW, potentially making this process economically attractive for scale-up, according to the authors.

As graphene is naturally occurring and shows a low toxicity profile, this could be an environmentally beneficial method to upcycle PW.

This process can expand the application of nanotechnology to the recycling of plastics and also expand their technological and social potentials of usage. For example, microplastics are a recent environmental issue observed due to the large consumption of plastic, which could be addressed by this technology.

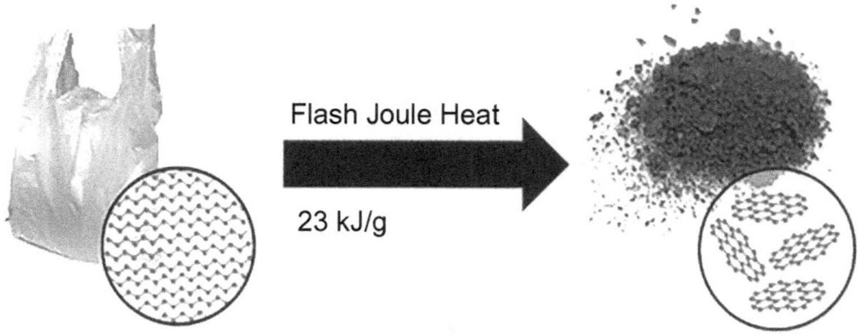

Fig. 8.6 The conceptual application of the flash Joule heat process to obtain nanostructured flash graphene from plastic waste. *(Reprinted from Algozeeb, W.A., Savas, P.E., Luong, D.X., Chen, W., Kittrell, C., Bhat, M., Shahsavari, R., Tour, J.M., 2020. Flash graphene from plastic waste. ACS Nano 14, 15595–15604, with permission from the American Chemical Society.)*

8.5 Conclusions

Nanotechnology is a development and application strategy that can promote the valorization of CO_2, biomass components, and recycling materials despite its unclear risks to the environment and public health.

Products such as catalysts, absorbents, nanotubes, and graphene can be obtained for application in several sectors like medicine, consumer products, energy, materials, and manufacturing. Furthermore, the application of nanotechnology to the recycling of plastics can expand its technological and social potentials.

References

Abdalla, S., Al-Marzouki, F., Al-Ghamdi, A.A., Abdel-Daiem, A., 2015. Different technical applications of carbon nanotubes. Nanoscale Res. Lett. 10, 358.

Algozeeb, W.A., Savas, P.E., Luong, D.X., Chen, W., Kittrell, C., Bhat, M., Shahsavari, R., Tour, J.M., 2020. Flash graphene from plastic waste. ACS Nano 14, 15595–15604.

Brunauer, S., Emmett, P.H., Teller, E., 1983. Adsorption of gases in multimolecular layers. J. Am. Chem. Soc. 60, 309–319.

Denton, R.D., Noyes, D., 2013. Method and System for Forming Carbon Nanotubes. Patent WO2013/090274A1. Patentscope. World Intellectual Property Organization.

Hassellöv, M., Kaegi, R., 2009. Analysis and characterization of manufactured nanoparticles in aquatic environments. In: Lead, J.R., Smith, E. (Eds.), Environmental and Human Impacts of Nanotechnology. Willey, Chichester.

Institute for Energy and Nuclear Research, 2021. Materials and Nanotechnology. https://www.ipen.br/portal_por/conteudo/documentos/201205161532040.8.materials_completo_baixa.pdf. (Accessed December 2021).

International Standard Organization, 2008. ISO/TS 27687:2008: Nanotechnologies: Terminology and Definitions for Nano-objects: Nanoparticle, Nanofiber and Nanoplate. ISO, Geneva.

Jiao, G.-J., Ma, J., Li, Y., Jin, D., Guo, Y., Zhao, J., Sun, R., 2021. Enhanced adsorption activity for phosphate removal by functional lignin-derived carbon-based adsorbent: optimization, performance and evaluation. Sci. Total Environ. 761, 143217.

Mulvaney, P., 2015. Nanoscience *vc* nanotechnology. ACS Nano 9, 2215–2217.

National Institute for Occupational Safety and Health, 2021. Nanotechnology. https://www.cdc.gov/niosh/topics/nanotech/default.html. (Accessed December 2021).

National Nanotechnology Initiative, 2021a. Improving our Energy Resources for Today and Tomorrow: Nanotechnology-Enabled Energy Solutions. https://www.nano.gov/node/1497. (Accessed December 2021).

National Nanotechnology Initiative, 2021b. Manufacturing at the Nanoscale. https://www.nano.gov/nanotech-101/what/manufacturing. (Accessed December 2021).

Quik, J.T.K., Meesters, J.A.J., Peijnenburg, W.J.G.M., Brand, W., Bleeker, A.J., 2020. Environmental risk assessment (ERA) of the application of nanoscience and nanotechnology in the food and feed chain. External Scientific Report. European Food Safety Authority. 133 pp.

Shuttleworth, P.S., De Bruyn, M., Parker, H.L., Hunt, A.J., Budarin, V.L., Matharu, A.S., Clark, J.H., 2014. Applications of nanoparticles in biomass conversion to chemicals and fuels. Green Chem. 16, 573–584.

Vance, M.E., Kuiken, T., Vejerano, E.P., McGinnis, S.P., Hochella Jr., M.F., Rejeski, D., Hull, M.S., 2015. Nanotechnology in the real world: redeveloping the nanomaterial consumer products inventory. Beilstein J. Nanotechnol. 6, 1769–1780.
Zdrojewicz, Z., Waracki, M., Bugaj, B., Pypno, D., Cabała, K., 2015. Medical applications of nanotechnology. Adv. Hyg. Exp. Med. 69, 1196–1204.

CHAPTER 9

Applied biotechnology for biomass

Biotechnological technology can be a very interesting alternative to produce some chemicals derived from renewable carbon (RC) raw materials, such as acids and alcohols, among other classes of organic molecules.

This chapter explores these possibilities and the main challenges to be overcome.

9.1 Introduction to industrial biotechnology

Industrial biotechnology is a multidisciplinary branch that is comprised of aspects of organic chemistry, analytical chemistry, biochemistry, chemical engineering, and microbiology to explore the biochemical routes of conversion for several classes of organic compounds.

The bioeconomy model (introduced in Chapter 1) demands a set of technologies to convert biomass into valuable products. From this necessity, industrial biotechnology is expected to achieve prominence when compared to other technology classes for some products, such as organic acids and alcohols, as observed by Nielsen et al. (2022).

As introduced in Chapter 4, the biochemical routes and their derived processes are divided into two categories:

- Enzymatic routes
- Microbial routes

Generally, the industrial application of biochemical processes for the conversion of RC raw materials involves:

- Fermentation technologies for the conversion step, e.g., yeasts.
- Biocatalysis technologies for the conversion step, e.g., immobilized enzymes
- Separation technologies for the downstream step, e.g., activated carbon, centrifugation, and chromatography

Fig. 9.1 depicts a typical flow diagram for fed-batch fermentation, whereas Fig. 9.2 depicts reactors for enzyme immobilization.

Renewable Carbon
https://doi.org/10.1016/B978-0-323-99735-5.00003-7

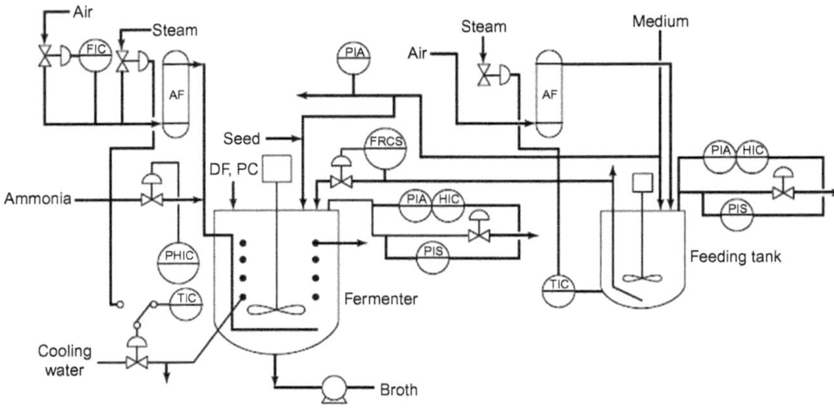

Fig. 9.1 Typical system for fed-batch fermentation. *AF*, air filter; *DF*, defoamer; *PC*, penicillin; *FIC*, flow indication control; *PHIC*, pH indication control; *TIC*, temperature indication control; *PIS*, pressure indication sum; *PIA*, pressure indication alarm; *HIC*, highest indication control; *FRCS*, flow record control sum. *(Credit: Sugimoto, M., 2002. Amino acids production process. In: Flickinger, M., Drew, S.W. (Eds.), Encyclopedia of Bioprocess Technology: Fermentation, Biocatalysis, and Bioseparation. vols. 1–5. Wiley, New York, pp. 89–100. Reprinted with permission from Wiley & Sons.)*

These classes of technologies and processes are well-explored in the book Ullmann's Encyclopedia of Industrial Chemistry, published by Wiley-VCH Verlag (https://onlinelibrary.wiley.com/doi/book/10.1002/14356007).

The use of this category of conversion technologies is more adequate for biomass as the RC source because the system works under mild conditions (i.e., pressure, temperature, and acidity/basicity) in an aqueous medium, reducing costs. Moreover, these aspects can be in line with the green chemistry principles, such as those related to renewable feedstock and reducing residue generation.

Nevertheless, because separation technologies are necessary to reach high purity of products, it can increase costs.

Table 9.1 presents some products obtained from different biomasses (substrates) and the biochemical process used for their production. When the substrates are pentose and hexose sugars, the possibility of obtaining them from the residual lignocellulosic biomass should be considered.

Some of the processes described in Table 9.1 are already well established (TRL9). However, for some compounds of great industrial interest, the use of biochemical processes is still in the proof-of-concept phase (TRL3). On the other hand, some compounds are being abandoned with a natural technological evolution.

The green chemistry challenge awards

The Green Chemistry Challenge Awards[a] promote the environmental and economic benefits of developing and using novel green chemicals in the

[a] https://www.epa.gov/greenchemistry/information-about-green-chemistry-challenge.

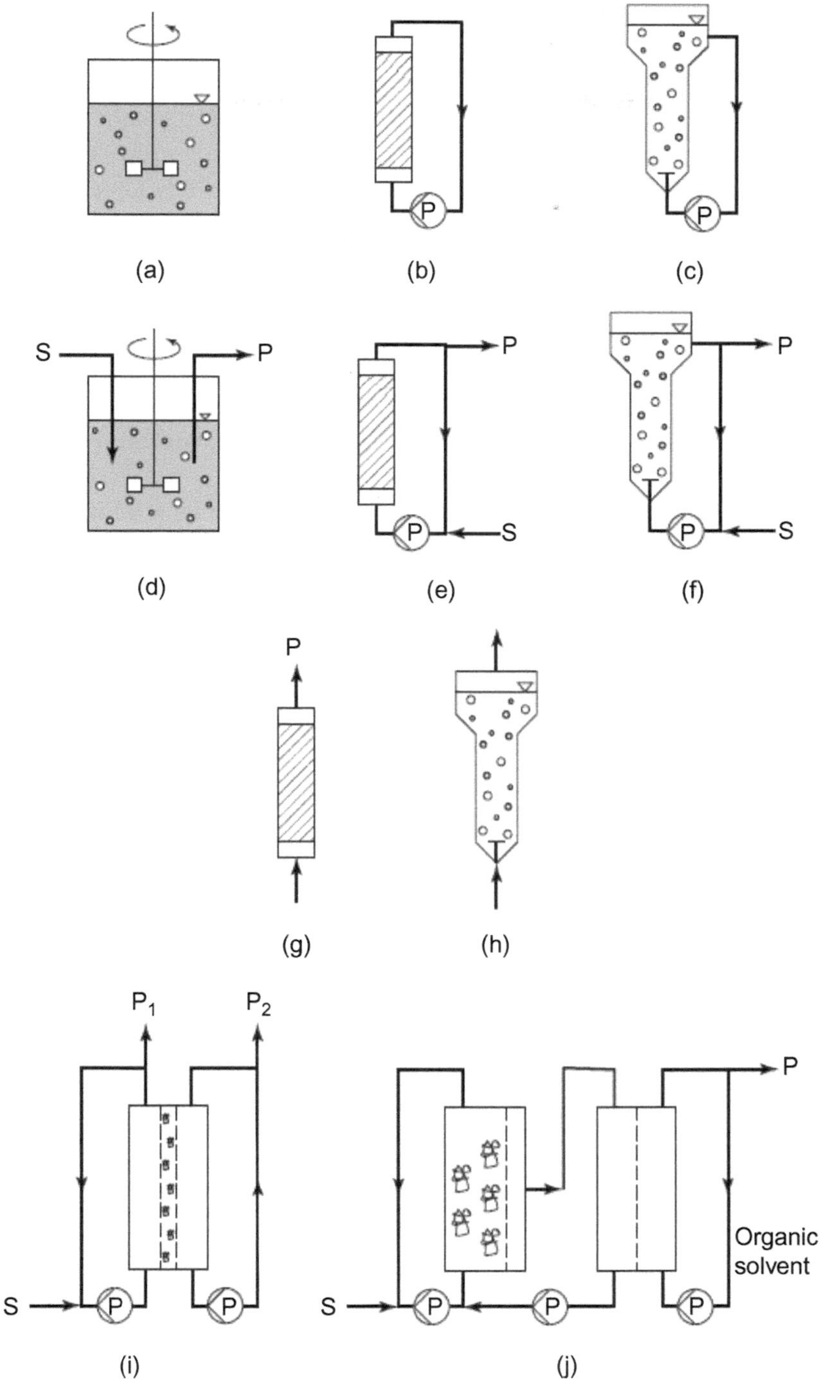
(a)
(b)
(c)
S
P
P
P
S
S
(d)
(e)
(f)
P
(g)
(h)
P1
P2
S
S
P
Organic
solvent
(i)
(j)

United States. These prestigious annual awards recognize chemical technologies that incorporate the principles of green chemistry in chemical design, manufacture, and use.

The Office of Chemical Safety and Pollution Prevention of the Environmental Protection Agency (EPA)[b] sponsors the Green Chemistry Challenge Awards in partnership with the American Chemical Society Green Chemistry Institute[c] and other members of the chemical community including industry, trade associations, academic institutions, and other government agencies.

A considerable number of biotechnologies can be noted among the winners, highlighting those related to renewable resources.

United States Environmental Protection Agency (2014): Fermentation using an engineered yeast produces β-farnesene, a hydrocarbon that can be used to make diesel fuel.

Innovation and benefits: renewable fuels are needed to help achieve global sustainability. Amyris took a step toward this goal by engineering yeast to make a chemical called farnesene instead of ethanol. Farnesene (Fig. 9.3) is a building block hydrocarbon that can be converted into a renewable, drop-in replacement for petroleum diesel without certain drawbacks of first-generation biofuels. Use of Amyris's renewable diesel may produce 82% less greenhouse gas (GHG) emissions than use of petroleum diesel.

Fig. 9.2 Reactors for immobilized enzymes. (A–C) Batch reactors with complete back mixing; (A) Stirred-tank reactor; (B) Fixed-bed reactor; (C) Fluidized-bed reactor. (D–F) Continuously operated reactors with complete back mixing. (G–H) Continuously operated reactors with plug-flow behavior. (I) Reactor with the enzyme immobilized in or on a membrane that may, at the same time, separate two phases, such as water and organic solvent. (J) Reactor with physically separated enzyme and organic solvent to prevent denaturation of the protein (the membrane reactor shown may be replaced by one of the types for immobilized enzymes). *S*, substrate; *P*, product. *(Credit: Kragl, U., Greiner, L., Wandre, C., 2002. Enzymes, immobilized, reactors. In: Flickinger, M., Drew, S.W. (Eds.), Encyclopedia of Bioprocess Technology: Fermentation, Biocatalysis, and Bioseparation. vols. 1–5. Wiley, New York, pp. 1064–1074. Reprinted with permission from Wiley & Sons.)*

[b] https://www.epa.gov/aboutepa/about-office-chemical-safety-and-pollution-prevention-ocspp#:~:text=OPPT%20manages%20programs%20under%20the,it%20gets%20into%20the%20environment.

[c] https://www.acs.org/content/acs/en/greenchemistry.html.

Table 9.1 Chemical compounds obtained through biochemical processes using different biomasses.

Chemical	Substrate	Type of process	Company
Acetic acid	Lignocellulose	Microbial process using *Lactobacillus casei* and *Clostridium thermoaceticum*	ZeaChem
Cis, *cis*-muconic acid	Glucose, alkanes	Enzymatic process	Amyris
Levulinic acid	Glucose or xylose	Microbial process and enzymatic process	The University of Maine Forest Bioproducts Research Institute (FBRI), Avantium, GFBiochemicals, Circa Group
Lactic acid	Glucose, sucrose, or lactose	Microbial process using *Lactobacillus*, *Sporolactobacillus*, *Enterococcus*, *Lactococcus*, *Bacillus*, *Streptococcus*, *Pediococcus*, or *Leuconostoc*	Purac, NatureWorks, Galactic, Henan Jindan, BBCA
Itaconic acid	Several sugars	Microbial process using *Aspergillus terreus* and *Aspergillus itaconicus*	Qingdao Kehai Biochemistry, Itaconix
Succinic acid	Several sugars	Microbial process using *Actinobacillus succinogenes* and *Mannheimia succiniciproducens*	BioAmber, Myriant, BASF/Purac, Reverdia (DSM/Roquette), PTT Chem/Mitsubishi CC
3-Hidroxypropionic acid	Several sugars or glycerol	Anaerobic fermentation using modified microorganisms	Cargill, Perstorp, OPXBio, DOW, Arkema
n-Butanol	Several sugars	Aerobic fermentation using modified microorganisms such as *Clostridium butylicum*	Cathay Industrial Biotech, Butamax, Butalco GmbH, Solvay (Cobalt assets)

Continued

Table 9.1 Chemical compounds obtained through biochemical processes using different biomasses—cont'd

Chemical	Substrate	Type of process	Company
iso-Butanol	Several sugars	Fermentation ABE	Butamax, Gevo
Caprolactam	Lysine	Microbial process	Highsun
Ethanol	CO_2, sucrose, cellulose, starch	Microbial process using microorganisms such as yeast *Saccharomyces cerevisiae*	Several ethanol producers around the world, highlighting producers from Brazil and United States
Ethylene glycol	Glucose or ethanol	Microbial process	India Glycols, Greencol Taiwan
Methane	Biomass, sewage, urban residues	Anaerobic digestion	Many companies around the world
Methanol	Pectin	Cultivation of *Clostridium butyricum*	Methanol BioMCN, Chemrec
n-Propanol	Glucose	Fermentative process using *Clostridium fallax* (postreaction: conversion to propylene)	Braskem
PHA (polyhydroxyalkanoates)	Hemicellulose, fatty acids, aromatics	Microbial process	Meridian Plastics, Tianjin Green Bioscience
Sorbitol	Glucose	Aerobic fermentation	Roquette, ADM
Xylitol	Xylose	Aerobic or anaerobic fermentation, or enzymatic conversion	DuPont

ABE = acetone-butanol-ethanol.

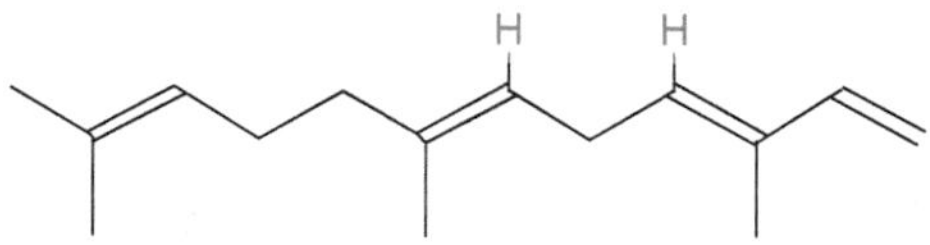

Fig. 9.3 The α-farnesene structure. *(Credit: author.)*

Summary of technology: transportation is second only to electricity generation as a source of GHG emissions in the United States. Gasoline consumption by automobiles is responsible for more than half of all GHG emissions from transportation, while other uses, such as diesel fuel for trucks, trains, or maritime use and jet fuel for aircraft, account for the remainder. Increases in fuel prices have led many countries, including the United States and Brazil, to begin modern large-scale production of biofuels. Locally produced biofuels have the potential to be a significant contributor to global sustainability, including providing local employment, ensuring access to energy resources, and reducing GHG emissions.

First-generation biofuels, notably ethanol and biodiesel (i.e., fatty acid methyl esters), suffer from limitations, such as limits on how much can be blended with gasoline or poor cold-weather performance. To address the known shortcomings of first-generation biofuels, Amyris developed an advanced renewable fuel compatible with the existing vehicle and distribution infrastructure that is now used in heavy-duty diesel engines and commercial aircraft.

Amyris used state-of-the-art strain engineering to make yeast that converts sugars into the hydrocarbon farnesene rather than ethanol. Farnesene can then be hydrogenated to farnesane, a renewable drop-in replacement for petroleum diesel and a blend-stock for jet fuel. A recent lifecycle analysis estimated an 82% reduction in GHG emissions for farnesane, compared with the EPA baseline fossil diesel—including indirect effects. Farnesane can also have land-use benefits for heavy-duty transportation: a hectare of land growing soybeans to produce traditional biodiesel generates enough fuel for a bus to travel about 600 miles. If the same land is instead used to grow sugarcane to make ethanol, a bus adapted to run on ethanol could travel about 4000 miles. However, if the sugarcane is used to produce farnesane, the unmodified diesel vehicle can travel about 5500 miles. Breakthroughs in converting lignocellulosic biomass to fermentable sugars will further increase this benefit.

Amyris has demonstrated industrial-scale production using its proprietary yeast strains to ferment sugars into renewable fuels that meet petroleum fuel specifications. Its renewable diesel, which has been approved by EPA for blends up to 35%, contains no sulfur or particulates, has a higher cetane number than diesel, and has improved low-temperature performance. In 2014, American Society for Testing and Materials (ASTM) revised its standard for jet fuel to include the use of renewable farnesane as a blending component in jet fuels for commercial aviation.

Analytical techniques

The quality control of biotechnological products and processes is conducted using some of the following analytical techniques:

- High Performance Liquid Chromatography (HPLC), using detectors such as ultraviolet-visible light (UV-VIS), mass spectrometry (MS), diode array detector (DAD), and refractive index detector (RID)—this is the most common technique for quantification.
- Size Exclusion Chromatography (SEC) with an enhanced light-scattering detector (ELSD)—for characterization.
- Gas Phase Chromatography (GC), using detectors such as flame ionization detector (FID) and mass spectrometry (MS)—for quantification.
- Spectrophotometry in the UV-VIS—for qualification.
- Fourier Transformed Infrared and Raman, by means probes—for characterization and qualification.
- OMICS: refers to a field of study in biological sciences that ends with -*omics*, such as genomics, transcriptomics, proteomics, or metabolomics; the ending *-ome* is used to address the objects of study of such fields, such as the genome, proteome, transcriptome, or metabolome, respectively (Vailati-Riboni et al., 2017). This field comprises a set of technologies based on MS and Nuclear Magnetic Resonance (NMR)—for structure resolution of biological systems.

A compendium of these techniques and their respective methods—except for OMICS—can be freely accessed, for example, from the National Renewable Energy Laboratory (NREL).[d] Other analytical techniques can be used according to the necessities; the book *Analytical Techniques and Methods for Biomass* (edited by Sílvio Vaz Jr. and published by Springer Nature)[e] will help.

[d] https://www.nrel.gov/bioenergy/laboratory-analytical-procedures.html.

[e] https://link.springer.com/book/10.1007/978-3-319-41414-0.

For OMICS, the book *Fundamentals of Advanced Omics Technologies: From Genes to Metabolites* (edited by Simó et al. and published by Elsevier)[f] is a valuable source.

9.2 Biotechnology applied to agroindustrial waste: A case study

Among the biomass residue treatment strategies is biochemical processing using enzymes (preferably those immobilized in a support) and microbes (e.g., fungi and bacteria). However, Vaz Jr. (2020) observed limitations related to purity and yields of the final products, which should be overcome to reach economic feasibility when considering operating (CAPEX) and capital (OPEX) costs (Issar and Navon, 2016; Scown et al., 2021). For example, the return of investment (ROI) can be applicable in the case of CAPEX (Eq. 9.1).

$$ROI = (Gain - Cost)/Cost \quad (9.1)$$

Enzymes have a high economic potential due to their applicability as biocatalysts, and this case study is based on the research conducted by Tacin et al. (2019) to apply lipases for industrial purposes.

Lipases are enzymes with a large number of applications at the industrial level, e.g., food, detergent, pharmaceutical, leather, textile, cosmetic, and paper industries. New producers and the development of bioprocess with high productivity are the major industrial challenges of these enzymes. Regarding the production process, oils from agroindustrial waste can emerge as efficient enzyme inductors, reducing the production cost. Moreover, this process can minimize environmental problems related to agroindustrial disposal. Thus, this work proposed to evaluate a new *Aspergillus* specie (Fig. 9.4) from an Amazon forest in Brazil as a lipase producer using oils from agroindustrial waste (grape seed oil [GSO] and cotton seed oil [CSO]) as an inductor. It used olive and soybean oil as the standard.

The microorganism metabolized all oils studied; however, the ones with a higher percentage of fatty acid esters (>80%), namely, soybean, olive, and GSO, promoted the highest lipase production. By comparing the activity of the lipase produced, GSO allowed an enzyme activity twofold higher than CSO. The enzyme produced using GSO was characterized and prepurified

[f] https://www.elsevier.com/books/fundamentals-of-advanced-omics-technologies-from-genes-to-metabolites/simo/978-0-444-62651-6.

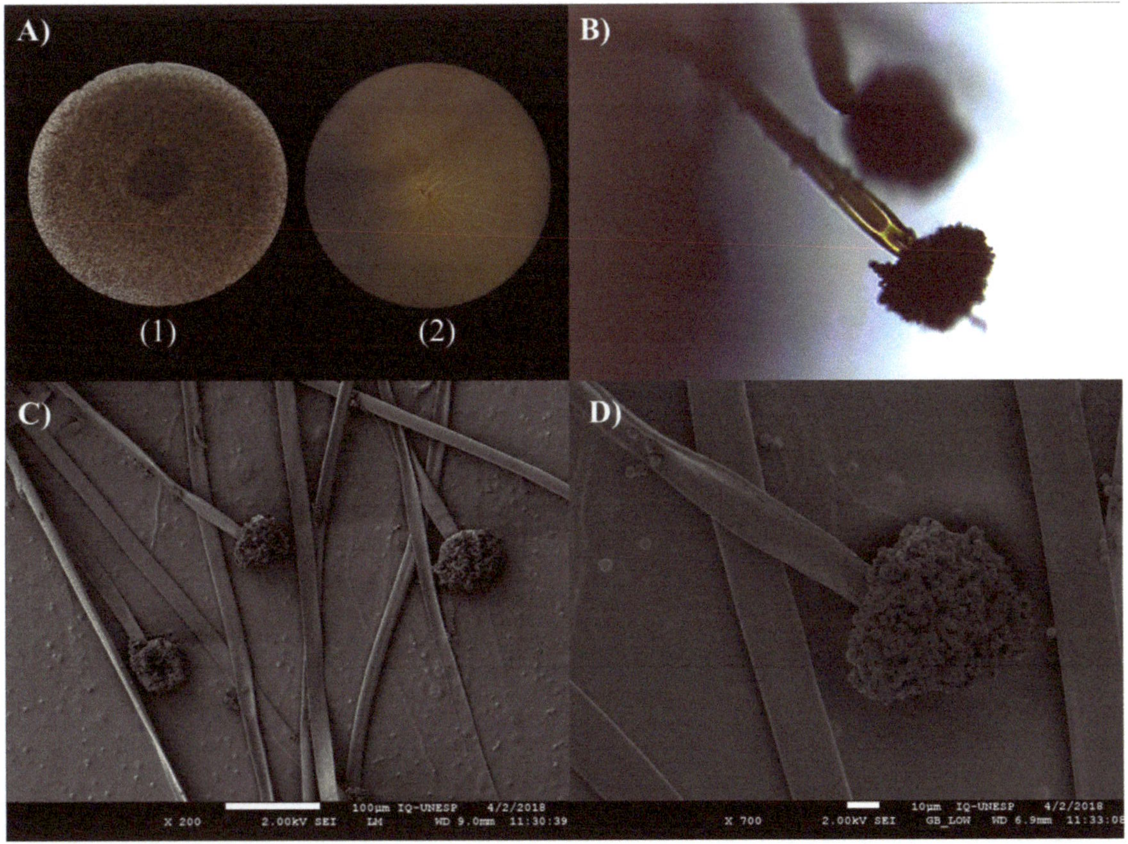

Fig. 9.4 Images of *Aspergillus* sp. DPUA 1727. (A) growth for 7 days in BDA (1,4-diamino-9,10-anthraquinone) plate; (B) conidiophore and conidia saw by optical microscopy; and (C and D) hyphae, conidiophores, dehydrated after treatment with 70% ethanol, and conidia seen by scanning electron microscopy (SEM). *(Credit: Tacin, M.V., Massi, F.P., Fungaro, M.H.P., Teixeira, M.F.S., de Paula, A.V., Santos-Ebinuma, V.D.C., 2019. Biotechnological valorization of oils from agro-industrial wastes to produce lipase using* Aspergillus *sp. from Amazon. Biocatal. Agric. Biotechnol. 17, 369–378. Reprinted with permission from Elsevier.)*

by precipitation, showing an optimum pH value of 7.5 and stability for pH between 7 and 7.5 values; an optimum temperature of 45°C and a thermal stability between 30°C and 50°C was achieved. In conclusion, a new isolate from the Amazon, *Aspergillus* sp., can produce lipase for industrial purposes using GSO. Moreover, other oils from agroindustrial waste with a high percentage of fatty acid esters can be used as an inductor for lipase production.

The use of an enzyme extracted from an Amazon *Aspergillus* specie that promotes waste biomass use confers a renewability appeal to the developed biotechnological process and its product, contributing to the greening of products and processes in accordance with principles 7 (renewable feedstock) and 9 (catalysis) of green chemistry.

9.3 Conclusions

Biotechnological technologies and processes can be used for RC conversion into valuable products. Some technologies are under development, and others are available for the industry.

However, limitations are observed related to purity and yields of the final products, which should be overcome to reach the economic feasibility—when considering operating and capital costs (CAPEX and OPEX).

References

Issar, G., Navon, L.R., 2016. Return on investment (ROI) calculations and CAPEX decisions. In: Operational Excellence. Management for Professionals. Springer, Cham.

Nielsen, J., Tillegreen, C.B., Petranovic, D., 2022. Innovation trends in industrial biotechnology. Trends in Biotechnology. In Press.

Scown, C., Baral, N.R., Yang, M., Vora, N., Huntington, T., 2021. Technoeconomic analysis for biofuels and bioproducts. Curr. Opin. Biotechnol. 67, 58–64.

Tacin, M.V., Massi, F.P., Fungaro, M.H.P., Teixeira, M.F.S., de Paula, A.V., Santos-Ebinuma, V.D.C., 2019. Biotechnological valorization of oils from agro-industrial wastes to produce lipase using *Aspergillus sp.* from Amazon. Biocatal. Agric. Biotechnol. 17, 369–378.

United States Environmental Protection Agency, 2014. Presidential Green Chemistry Challenge: 2014—Small Business Award. EPA. https://www.epa.gov/greenchemistry/presidential-green-chemistry-challenge-2014-small-business-award.

Vailati-Riboni, M., Palombo, V., Loor, J.J., 2017. What are omics sciences? In: Ametaj, B. (Ed.), Periparturient Diseases of Dairy Cows. Springer, Cham.

Vaz Jr., S., 2020. Treatment of Agroindustrial Biomass Residues – A Sustainable Approach. Springer Nature, Cham.

CHAPTER 10

Applied thermochemistry for biomass and recycling materials

Thermochemistry, the branch of chemical sciences that makes use of thermal energy to perform reactions and processes, can be very useful to treat biomass and plastic residues and promote new products made from them.

This chapter explores the main thermochemical processes for renewable carbon (RC) raw materials.

10.1 Introduction to industrial thermochemistry

Thermochemistry for industrial purposes applied to RC raw materials comprises a set of technologies and processes based on different temperature ranges and operational conditions, as introduced in Chapter 3. It can be very useful to treat biomass and plastic residues to reduce environmental impacts and develop new products.

Table 10.1 describes the main characteristics and differences of the thermochemical processes for RC, in order to identify those more suitable for biomass - especially lignocellulosic biomass, e.g., wood.

We can observe from Table 10.1 that the energy cost can be the main parameter to consider for the operational expenditure (OPEX) in an economic feasibility assessment. Energy is a very expensive input for the industry according to its origin, which means hydroelectricity is cheaper than thermoelectricity because the last uses fossil fuels (i.e., coal, diesel, and natural gas) whereas the former uses surface water moved by gravity. Table 10.2 presents the main fuels - used as energy source - associated with thermochemical processes.

In a general way, it is expected that a fuel has the following properties (El-Mahollawy and Habik, 2002):

- High energy density (content)
- High heat of combustion (release)
- Good thermal stability (storage)
- Low vapor pressure (volatility)
- Nontoxicity (environmental impact)

Renewable Carbon
https://doi.org/10.1016/B978-0-323-99735-5.00006-2

Table 10.1 Main operational conditions characteristics of the thermochemical processes for RC raw materials, especially for biomass.

Thermochemical process	Temperature range (°C)	Heating rate	Pressure	Residence time	Primary product(s)
Combustion	>300	Slow–fast	Atmospheric–elevated	Seconds–minutes	Heat
Slow pyrolysis	350–800	Slow ($<10°C\,min^{-1}$)	Atmospheric	Hours–days	Char
Torrefaction	200–300	Slow ($<10°C\,min^{-1}$)	Atmospheric	Minutes–hours	Stabilized, friable biomass
Fast pyrolysis	400–600	Very fast ($\sim1000°C\,s^{-1}$)	Vacuum–atmospheric	Seconds	Bio-oil
Flash pyrolysis	300–800	Fast	Elevated	Minutes	Biocarbon/char
Gasification	700–1500	Moderate–very fast	Atmospheric–elevated	Seconds–minutes	Syngas

Table 10.2 Classification of chemical fuels by their physical state and availability.

Naturally available (reference)	Synthetically produced
Solid	–
Coal	Charcoal
Wood	Inorganic solid waste
Vegetation	–
Organic solid waste	–
Liquid	
Crude oil	Syncrude
Biological oils	Petroleum distillates
Fuel plants	Alcohols
–	Colloidal fuels
–	Benzene
Gas	
Natural gas	Natural gas
Marsh gas[a]	Hydrogen
Biogas	Methane
–	Propane
–	Coal gasification

[a]Moisture of methane, hydrogen sulfide, carbon dioxide, and trace of phosphine produced naturally by means the organic matter decomposition by biochemical processes such as methanogenesis.
Source: Adapted from Keating, E. 2007. Applied Combustion. Taylor & Francis, New York. Reproduced with permission from Marcel Dekker.

Table 10.3 describes the calorific values[a] of the main fuels used today by modern society. The majority of them are carbon-derived substances; however, they come from fossil sources and not from a RC source, which should be reconsidereded in order to improve the sustainability of energy generation, taking into account the reuse of a huge variaty of residues.

Thermochemistry processes can be used to generate several products, and Fig. 10.1 describes the potential of products derived from biomass by means of physicochemical, biological, and thermochemical conversions.

On the other hand, Fig. 10.2 describes the potential of products derived from plastics in electronic devices (e-waste) by means of thermochemical conversion and other treatment strategies.

Products like biochar, crude bio-oil, and biosyngas can be strategically used to get functional materials, chemicals, and liquid and gas fuels. These bioproducts can be explored in a biorefinery facility, as proposed by Kamm et al. (2006) and Vaz Jr. (2018).

[a] Calorific value (CV) is the amount of energy released in the form of heat when a fuel is completely oxidized, producing CO_2. It is usually expressed as a unit of energy per unit of mass.

Table 10.3 Calorific value (CV) of fuels.

Fuel	Calorific value (kcal kg^{-1})	Uses
Wood	2500–4000	Heating, cooking
Carbon (coal)	4000–9000	Heating
Diesel oil	10,325	Transportation
Gasoline	10,400	Transportation
Butane	10,900	Heating
Petroleum	11,000	Fuels production
Liquefied petroleum gas (LPG)	11,000	Transport, cooking
Natural gas (methane)	11,500	Heating, cooking
Hydrogen	28,700	Transportation

Applications presented in Figs. 10.1 and 10.2 have a strong potential to create wide "permeable" value chains.

Sustainable chemistry, the chemistry developed and executed according to sustainability components (discussed in Part 3), can be promoted, among other technological strategies, by the use of thermochemical conversion to obtain chemicals, materials, and energy of second and third generations, considering the carbon footprint.[b] A very interesting alliance between thermochemistry and CCU strategy (Chapter 7) coming from the postcombustion capture of CO_2 from coal burning and its conversion into carbonates for several usages.

Chemical recycling

Chemical recycling is a tertiary-type recycling (American Society for Testing and Materials, 2000) that involves pyrolysis, solvolysis, and gasification technologies, as described in Table 10.4. It is a valuable strategy to reduce the negative impacts from plastic wastes and general recycling materials.

Although solvolysis is not a thermochemical process, it can be thought of as an organic-inorganic strategy that depends on the applied solvent to reach the best results.

Analytical techniques

Analytical techniques are paramount to monitoring and controlling processes and products obtained by thermochemistry conversion routes. The main techniques are as follows:

- Ash Content: the incombustible component remaining after a sample is completely burned, which can be composed of inorganic salts, such as metallic oxides. It is a gravimetric technique that measures the relative mass difference.

[b] https://unesdoc.unesco.org/ark:/48223/pf0000217181.

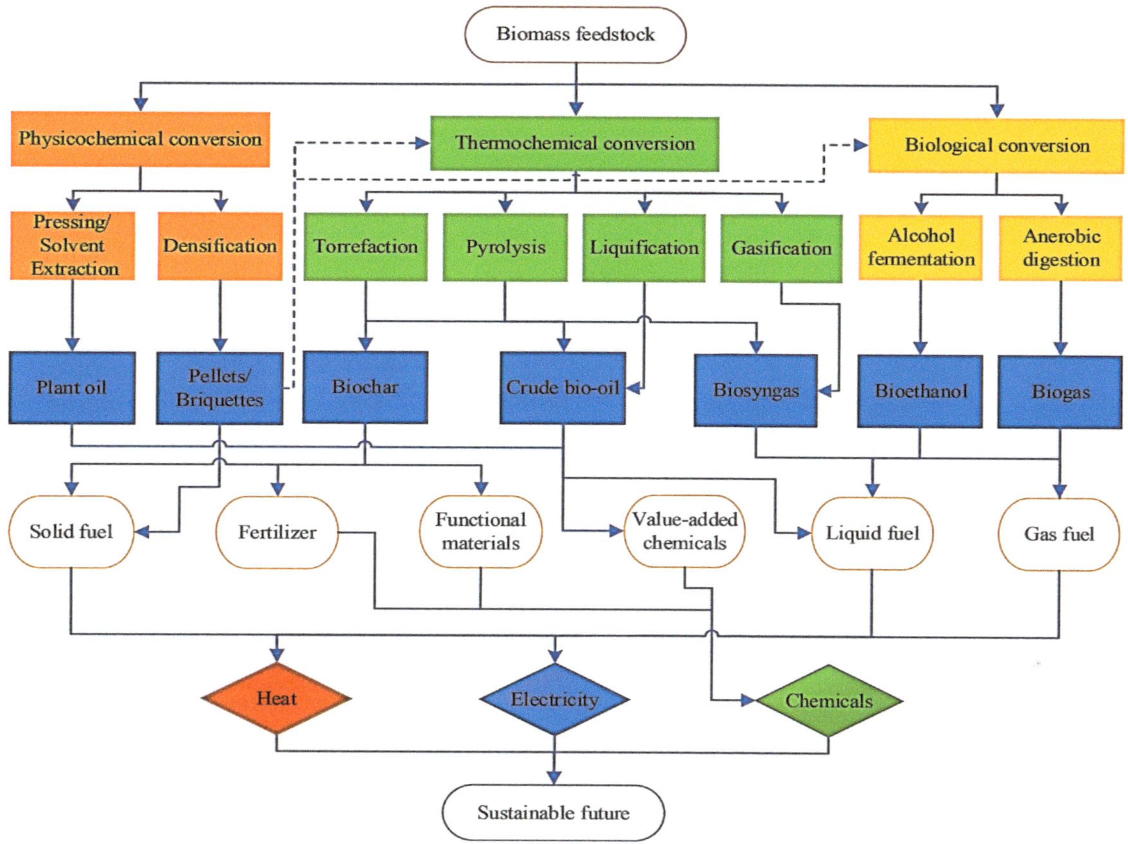

Fig. 10.1 Major biomass conversion pathways into renewable energy and chemicals. Thermochemical conversion is highlighted in *green* boxes. *(Credit: Kang, K., Klinghoffer, N.B., El Ghamrawy, I., Berruti, F., 2021. Thermochemical conversion of agroforestry biomass and solid waste using decentralized and mobile systems for renewable energy and products. Renew. Sust. Energ. Rev. 149, 111372. Reproduced with permission from Elsevier.)*

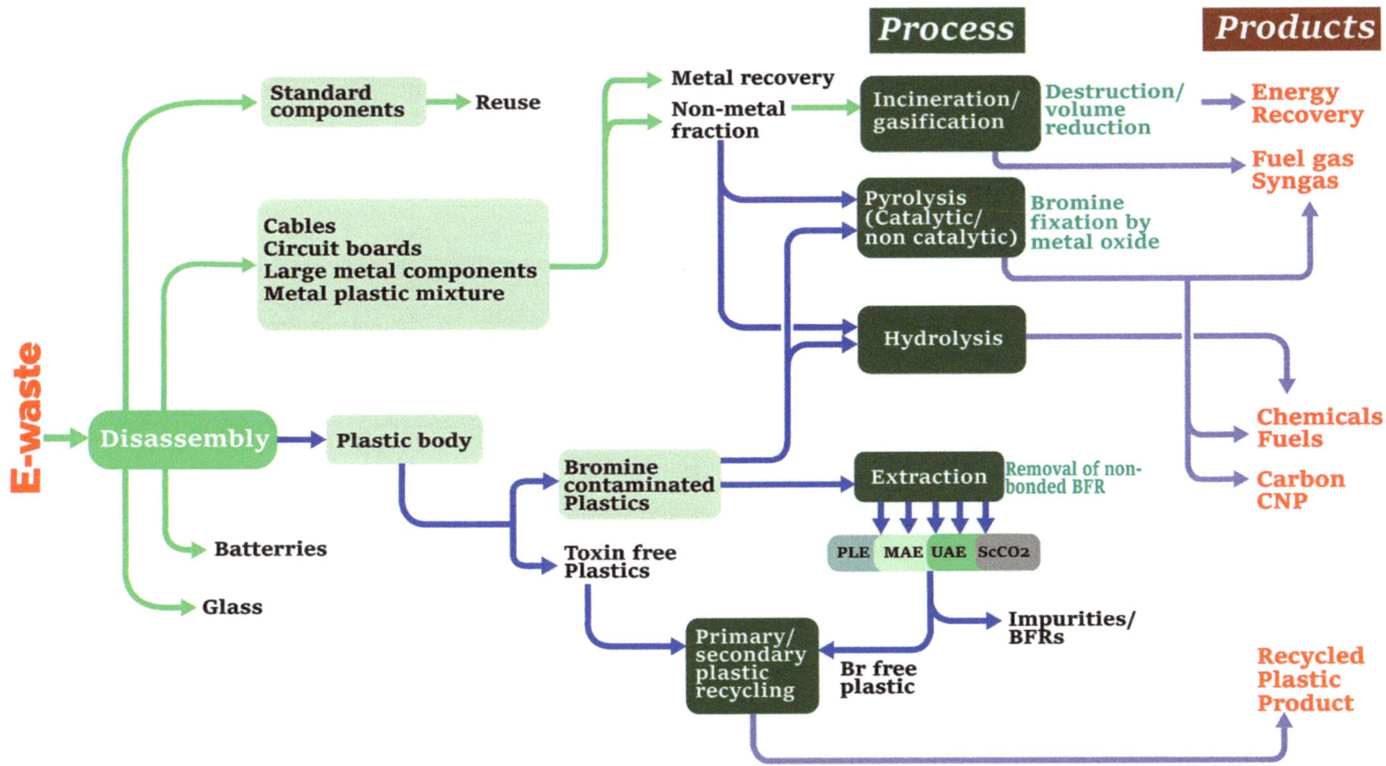

Fig. 10.2 Simplified flow chart of e-waste plastics from e-waste stream to product stream. *BFRs*, brominated flame-retardants; *PLE*, pressurized liquid extraction; *MAE*, microwave assisted extraction; *UAE*, ultrasonic assisted extraction; *ScCO$_2$*, supercritical carbon dioxide. *(Credit: Das, P., Gabriel, J.-C. P., Tay, C.Y., Lee, J.-M.L., 2021. Value-added products from thermochemical treatments of contaminated e-waste plastics. Chemosphere 269, 129409. Reprinted with permission from Elsevier.)*

Table 10.4 Summary of different chemical recycling approaches.

Recycling method	Process description	Polymer type	Advantages	Challenges
Pyrolysis	Degradation of polymeric chains in presence of the inert atmosphere	All plastics	Relatively no significant release of toxic gases Produce electricity and heat Broad distribution of solid, liquid, and gaseous products	Complex chemistry Requires high volumes to be cost-effective Heteroatoms reduce product quality
Solvolysis	Solvent medium is used to break the plastic waste into low molecular weight products	PET, PLA, PC	Manageable and effective Extensive reactions Solvent to depolymerization	Long time for complete depolymerization Cannot handle high content additives Difficult to separate feedstock impurities
Gasification	Partial oxidation process (using air or steam) for the conversion of plastic wastes into gases	All plastics	Polymer separation into different categories is not necessary Produce electricity and heat Heat to depolymerization	Release of toxic gases Lacks technology for gasification of mixed plastic waste Cannot be used in monomer production Tar formation

PET, poly(ethylene) terephthalate; *PLA*, polylactic acid; *PC*, polycarbonate.
Adapted from Jiang, J., Shi, K., Zhang, X., Yu, K., Zhang, H., He, J., Ju, Y., Liu, J., 2021. From plastic waste to wealth using chemical recycling: a review. J. Environ. Chem. Eng. 106867 (in press). Reprinted with permission from Elsevier.

- Pyrolysis Gas Phase Chromatography Mass Spectrometry (Py-GC-MS): can be used to characterize most materials including insoluble and complex materials at trace levels often without any sample pretreatment, e.g., polymers, plastics, rubber, paints, dyes, resins, coatings, cellulose, wood, textiles, oils, etc.
- Thermogravimetric Analysis (TGA): observes the thermal decomposition of the sample by means of mass loss according to the temperature increase.
- Differential Scanning Calorimetry (DSC): for determining the glass transition temperature of the sample—a physicochemical property; the difference in the amount of heat required to increase the temperature of a sample and a reference material is measured as a function of temperature.
- Fourier Transformed Infrared Absorption Spectroscopy (FTIR): for structural elucidation by infrared absorption producing absorption bands based on vibrational modes, which is characteristic for certain chemical groups according their polarity.
- Scanning Electron Microscopy (SEM) combined with the identification of chemical elements present in the sample surface by Energy-dispersive X-ray Spectroscopy (EDS): for the observation of the morphology of the sample surface—to determine the physicochemical property—and for the determination of its chemical composition.
- Transmission Electron Microscopy (TEM): as with SEM, it is a microscopy technique that permits higher magnifications than SEM and also shows the crystallographic structure and composition of a material.
- X-ray Spectroscopy (XRS): used for the elemental, chemical, crystalline, structural, and dynamic analysis of a broad range of materials.
- Raman Spectroscopy: a nondestructive chemical analysis technique that provides detailed information about chemical structure, phase, and polymorphy, crystallinity, and molecular interactions. It is based on the interaction of infrared radiation with the chemical bonds within a material and is very similar to FTIR.

A compendium of these techniques and their respective methods can be freely accessed from the National Renewable Energy Laboratory (NREL).[c] Books such as *Analytical Techniques and Methods for Biomass* (edited by Sílvio Vaz, Jr., published by Springer Nature)[d] can be used as a guide for the chemical analysis of RC-derived products.

[c] https://www.nrel.gov/bioenergy/laboratory-analytical-procedures.html.
[d] https://link.springer.com/book/10.1007/978-3-319-41414-0.

As observed in Chapter 8, the analytical approach and, consequently, the best techniques for use will depend on the sample characteristics (e.g., physical state). Eventually, other techniques not listed here can be used.

10.2 Thermochemistry applied to agroindustrial biomass residues: A case study

This case study is based on Solarte-Toro et al. (2021) regarding the comparison of the gasification, pyrolysis, and combustion of woody biomass from a technoeconomic perspective.

Pinus patula was used as raw material to be upgraded into electricity, syngas, and bio-oil through combustion, gasification, and pyrolysis, respectively. A technoeconomic comparison was carried out to elucidate the best process to convert this raw material at a low scale. For this, a processing scale of 250 t day^{-1} was considered. The Aspen Plus software v.9.0[e] was used to perform the simulation of the processes. The chemical characterization, proximate analysis, and heating value reported by García-Velásquez and Cardona (2019) were used.

The simulated thermochemical conversion processes were composed of three different stages: biomass pretreatment, thermochemical upgrading process, and power generation section. In the pretreatment stage, the raw material was dried and milled until it reached an appropriate particle size. Then, the thermochemical conversion process was done to obtain the main products. The process flow diagram of the gasification, combustion, and pyrolysis processes are presented in Figs. 10.3, 10.4, and 10.5. The gasification process involved the production of syngas, tars, biochar, and ash. In this process, biochar and ash were evaluated as possible value-added products to be commercialized. On the other hand, the syngas produced was used as fuel to generate power in a gas engine. The gasification process was carried out at 800°C using pressurized air as an oxidizing agent. Also, the syngas combustion was designed to reach the adiabatic flame temperature at the outlet of the gas engine. Subsequently, the process exhaust gases were used to preheat the inlet air to the gasification process. The intermediate pyrolysis process flow diagram is presented in Fig. 10.4. This process involved the production of bio-oil, biochar, and electricity. The pyrolysis process was carried out at 400°C. Nitrogen was used to guarantee a nonoxidizing atmosphere in the

[e] https://www.aspentech.com/en/resources/press-releases/aspenonert-version-9-software-delivers-the-innovation-that-creates-intelligent-insigh2147498788.

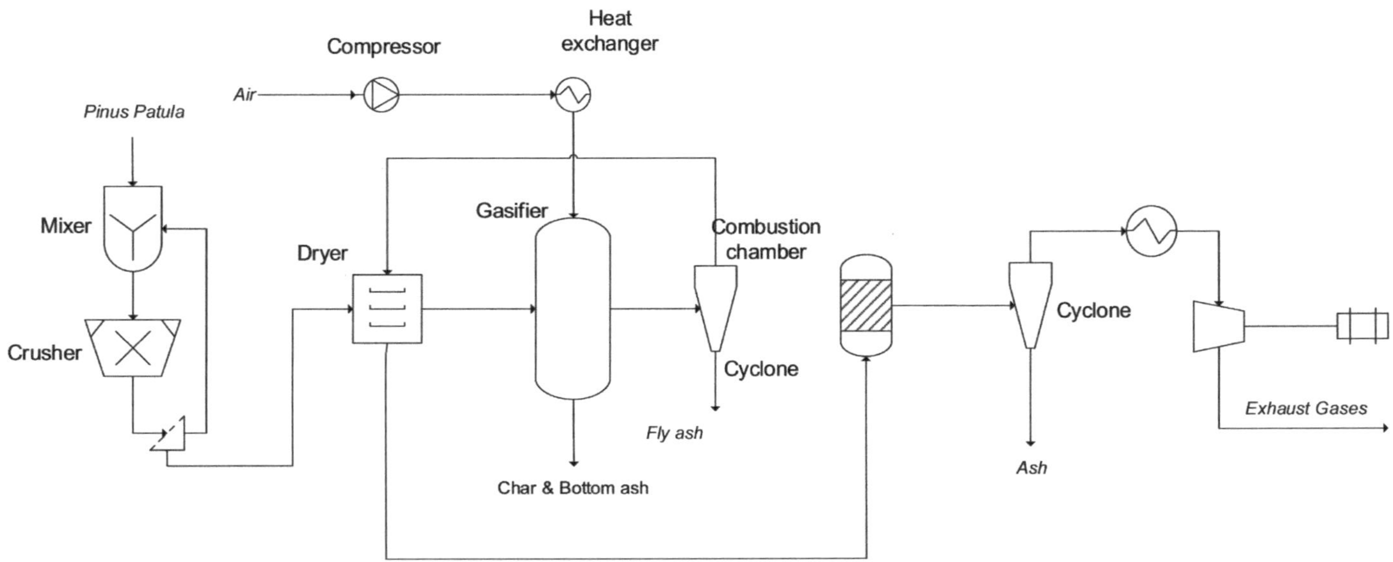

Fig. 10.3 Process flow diagram of woody biomass gasification process for power generation. *(Credit: Solarte-Toro, J., González-Agirre, J.A., Giraldo, J.A.P., Alzate, C.A.C., 2021. Thermochemical processing of woody biomass: a review focused on energy-driven applications and catalytic upgrading. Renew. Sust. Energ. Rev. 136, 110376. Reprinted with permission from Elsevier.)*

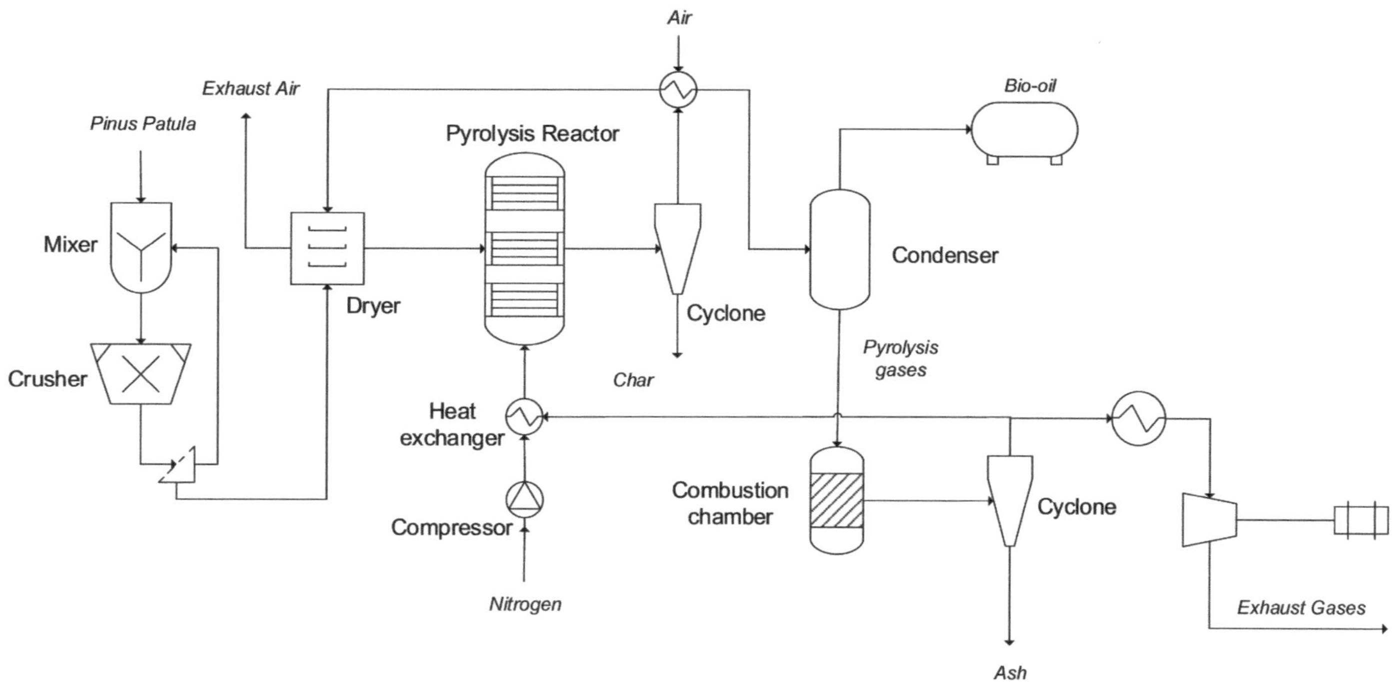

Fig. 10.4 Process flow diagram of woody biomass pyrolysis process for power generation. *(Credit: Solarte-Toro, J., González-Agirre, J.A., Giraldo, J.A.P., Alzate, C.A.C., 2021. Thermochemical processing of woody biomass: a review focused on energy-driven applications and catalytic upgrading. Renew. Sust. Energ. Rev. 136, 110376. Reprinted with permission from Elsevier.)*

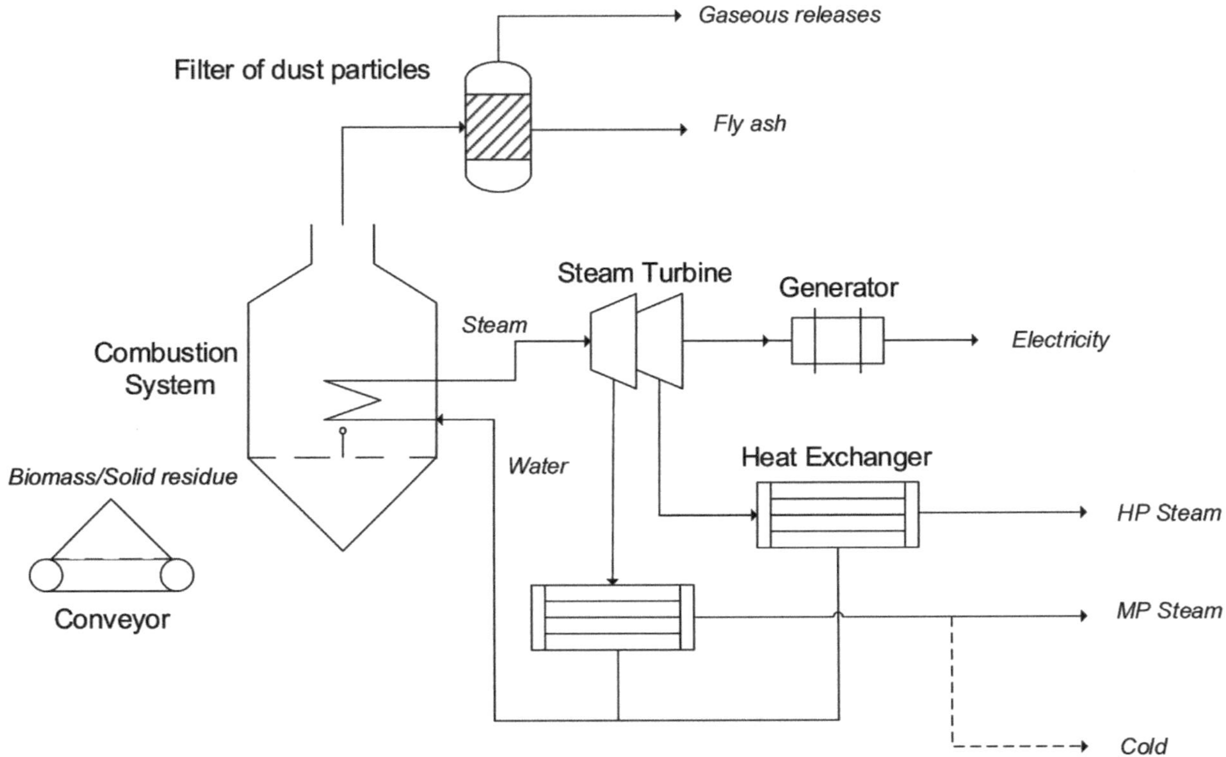

Fig. 10.5 Process flow diagram of the woody biomass combustion process for power generation. *HP*, high pressure; MP, medium pressure. *(Credit: Solarte-Toro, J., González-Agirre, J.A., Giraldo, J.A.P., Alzate, C.A.C., 2021. Thermochemical processing of woody biomass: a review focused on energy-driven applications and catalytic upgrading. Renew. Sust. Energ. Rev. 136, 110376. Reprinted with permission from Elsevier.)*

pyrolysis reactor. In this process, the pyrolysis gases (i.e., CO, H_2, CH_4, and CO_2) were combusted to produce electricity.

Finally, the combustion process was simulated using the flowsheet presented in Fig. 10.5. This process was designed considering an oxygen excess of 25% to ensure complete combustion, a temperature of 900°C, and pressure of 1 atm.

Results

The carbon conversion efficiency was calculated for the gasification and pyrolysis cases. A 94.98% value was obtained in the gasification case; in the same way, the cold gas efficiency of the gasification process was calculated as 72.95%; the carbon conversion efficiency of the gasification process was higher than in the case of the pyrolysis process (75.63%). The combustion process had the highest process efficiency (74.02%) due to the maximum degradation rate of the materials that compose the woody raw material demonstrating that *P. patula* has a high potential for direct combustion.

The economic assessment of the thermochemical alternatives to upgrade woody biomass was done in terms of CAPEX and OPEX separately. CAPEX values determined were US \$936,630 (combustion), US \$2,364,530 (gasification), and US \$7,657,175 (pyrolysis). OPEX values determined were US \$121,750 $year^{-1}$ (combustion), US \$454,562 $year^{-1}$ (gasification), and US \$1,164,069 $year^{-1}$ (pyrolysis).

Although this case study is a simulation, it can provide parameters for investments to support the application of this category of technologies.

10.3 Thermochemistry applied to recycling materials: A case study

This case study is based on the research of Thunman et al. (2019) regarding thermochemical plastic recycling according to a circular economy. Experiments were conducted in reactors. It is an application of chemical recycling ("Chemical recycling" section).

Progression toward using plastic materials in a circular way requires a technology that treats any type of plastic waste (PW)—sorted or unsorted—and produces plastics of the same quality as the original. This would close the material cycle, transitioning from flows directed according to the waste hierarchy to circular use, i.e., the fraction that currently ends up in energy recovery or disposal becomes the feedstock for new plastics. This can be achieved by thermochemical recycling, which entails theoretically unlimited recycling of any plastic material (mixed or sorted), where the focus is on recovering the building blocks of the plastic materials. This solution can be applied in any part of the world regardless of the availabilities of advanced sorting systems, and, more importantly, it creates an economic driver for the collection and logistics of PW. Thermochemical recycling becomes an upgrading bridge between PW streams and a phase of new use. In fact, thermochemical recycling maximizes the increase in value of the carbon atom in the PW compared to that obtained using any other existing recycling route, such as mechanical recycling, where the new use phase is ensured, albeit with a product of lower quality.

Fig. 10.6 presents the thermochemical conversion strategy.

Results

From Fig. 10.6, Route A is preferable from a thermodynamic point of view as it preserves the structures of the existing molecules. However, the direct formation of valuable monomers is dependent upon the nature and composition of the feedstock applied. In contrast, Routes B and C are less fuel-dependent and confer fuel flexibility upon the overall process.

Fig. 10.7 illustrates the application of the strategies presented in Fig. 10.6 to reach 100% carbon recovery.

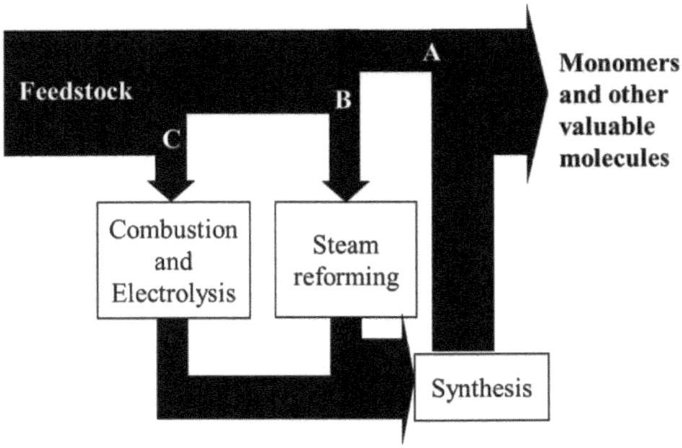

Fig. 10.6 Flows of carbon in the three thermal recycling routes for carbon recovery. *(Credit: Thunman, H., Vilches, T.B., Seeman, M., Maric, J., Lela, I.C., Pissot, S., Nguyen, H.N.T., 2019. Circular use of plastics-transformation of existing petrochemical clusters into thermochemical recycling plants with 100% plastics recovery. Sustain. Mater. Technol. 22: e00124. Reprinted with permission from Elsevier.)*

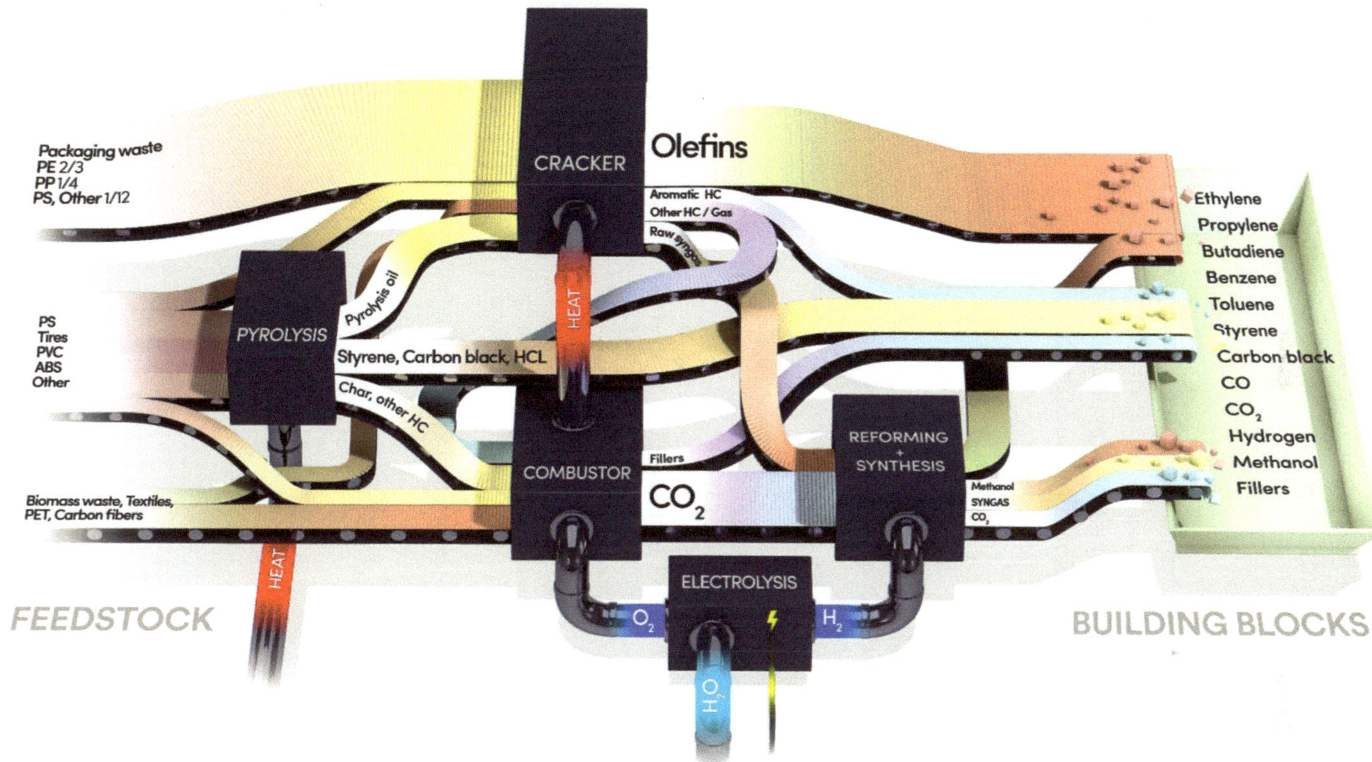

Fig. 10.7 Schematic of a technical solution for closing the material cycle of plastics using thermochemical recycling, which can handle any type of plastic waste (sorted or mixed) with close to 100% carbon recovery. *ABS*, acrylonitrile butadiene styrene; *PET*, poly(ethylene terephthalate); *PE*, polythene; *PP*, polypropylene; *HC*, hydrocarbon; *PVC*, polyvinyl chloride; *PS*, polystyrene. *(Credit: Thunman, H., Vilches, T.B., Seeman, M., Maric, J., Lela, I.C., Pissot, S., Nguyen, H.N.T., 2019. Circular use of plastics-transformation of existing petrochemical clusters into thermochemical recycling plants with 100% plastics recovery. Sustain. Mater. Technol. 22, e00124. Reprinted with permission from Elsevier.)*

As presented by the authors, thermochemical conversion is a smart strategy to possibility transform PW into valuable products, considering the carbon footprint (to be treated in the Part 3).

10.4 Conclusions

Thermochemical conversion, despite the high demand of energy to make it an ongoing process, can be seen as a smart set of technologies to treat and reuse biomass and PWs. For example, it is expected to reach 100% carbon recovery. Then RC can be explored to produce chemicals, materials, and energy of the second and third generation of sustainable chemistry.

References

American Society for Testing and Materials, 2000. Standard Guide for Development of STM Standards Relating to Recycling and Use of Recycled Plastics. ASTM D5033. ASTM International, West Conshohocken.

El-Mahollawy, F., Habik, S.E.-D., 2002. Fundamentals and Technology of Combustion. Elsevier, Amsterdam.

García-Velásquez, C.A., Cardona, C.A., 2019. Comparison of the biochemical and thermochemical routes for bioenergy production: a techno-economic (TEA), energetic and environmental assessment. Energy 172, 232–242.

Kamm, B., Gruber, P.R., Kamm, M. (Eds.), 2006. Biorefineries – Industrial Processes and Products. Status Quo and Future Directions. vol. 1. Wiley-VCH, Weinheim.

Solarte-Toro, J., González-Agirre, J.A., Giraldo, J.A.P., Alzate, C.A.C., 2021. Thermochemical processing of woody biomass: a review focused on energy-driven applications and catalytic upgrading. Renew. Sust. Energ. Rev. 136, 110376.

Thunman, H., Vilches, T.B., Seeman, M., Maric, J., Lela, I.C., Pissot, S., Nguyen, H.N.T., 2019. Circular use of plastics-transformation of existing petrochemical clusters into thermochemical recycling plants with 100% plastics recovery. Sustain. Mater. Technol. 22, e00124.

Vaz Jr., S. (Ed.), 2018. Biomass and Green Chemistry – Building a Renewable Pathway. Springer Nature, Cham.

CHAPTER 11

Catalysis for carbon dioxide and biomass

The catalytic conversion of renewable carbon (RC) raw materials is expected to boost the use of this type of feedstock by the industry to obtain several chemicals.

To explore opportunities and challenges related to this technological strategy, this chapter presents and discusses fundamental and applied aspects.

11.1 Introduction to industrial catalysis

The conversion of biomass and CO_2 into chemicals is, mostly, unfavorable from both the thermodynamic and kinetic point of view, which leads to the need for catalysts to perform the related reactions. To differ it from biocatalysis using enzymes, it is defined as chemocatalysis.

According the International Union of Pure and Applied Chemistry (2021a), a catalyst is a substance that increases the rate of a reaction without modifying the overall standard Gibbs energy change in the reaction, a process called catalysis. The catalyst is both a reactant and product of the reaction. The words *catalyst* and *catalysis* should not be used when the added substance reduces the rate of reaction. Catalysis can be classified as homogeneous and heterogeneous; in the first, only one phase is involved, and in the second, the reaction occurs at or near an interface between phases.

Catalysis brought about by one of the products of a reaction is called autocatalysis. Catalysis brought about by a group of reactant molecules is called intramolecular catalysis. The term catalysis is also often used when the substance is consumed in the reaction, for example, base-catalyzed hydrolysis of esters. Strictly, such a substance should be called an activator (International Union of Pure and Applied Chemistry, 2021a).

Catalysts allow the reaction to proceed along a path at a lower activation energy (Fig. 11.1), which leads to an increase in reaction rate and a decrease in reaction time.

Renewable Carbon
https://doi.org/10.1016/B978-0-323-99735-5.00012-8

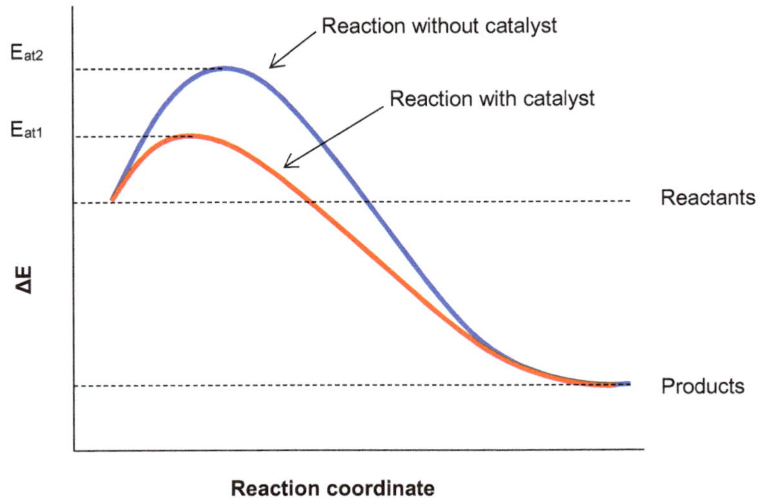

Fig. 11.1 Comparison between activation energies for a reaction with the presence of catalyst (Eat_1) and without the presence of catalyst (Eat_2). *(Credit: author.)*

The reaction rate is paramount to be understood to best explore the catalytic conversion.

Again, according the International Union of Pure and Applied Chemistry (2021b), for the general chemical reaction described in Eq. 11.1:

$$aA + bB \rightarrow pP + qQ + \ldots aA + bB \rightarrow pP + qQ + \ldots \quad (11.1)$$

Occurring under constant-volume conditions, without an appreciable build-up of reaction intermediates, the reaction rate ν is defined by Eq. 11.2:

$$\nu = 1/q \times d[Q]/dt \quad (11.2)$$

where symbols placed inside the square brackets denote amount (or amount of substance) of concentrations (conventionally expressed in units of $mol\,dm^{-3}$). The reaction rate differs from the rate of increase of concentration of a product P by a constant factor (the reciprocal of its coefficient in the stoichiometric equation) and from the rate of decrease of concentration of the reactant A by α^{-1}.

More generally, it is recommended that, instead, the terms *rate of disappearance* or *rate of consumption* of A (i.e., $-d[A]/dt$, the rate of decrease of concentration of A) or *rate of appearance* of P (i.e., $d[P]/dt$, the rate of increase of concentration of product P) be used, depending on the concentration change of the particular chemical species that is actually observed. In some cases, reference to the chemical flux observed may be more appropriate.

The symbol ν (without lettered subscript) should be used only for reaction rate; ν with a lettered subscript (e.g., νA) refers to a rate of appearance or rate of disappearance (e.g., of the chemical species A).

The calculations of mass (Eq. 11.3) and molar (Eq. 11.4) yields for the target products are of fundamental importance to verify the efficiency of the idealized process.

$$Y_{mas}\ (\%) = m_p/m_r \times 100 \tag{11.3}$$

where m_p is the mass of product formed (g, kg) and m_r is the mass of consumed reactant (g, kg).

$$Y_{mol}\ (\%) = n_p/n_r \times 100 \tag{11.4}$$

where n_p is the number of moles of product formed and n_r is the number of moles of consumed reagent.

Conversion (Eq. 11.5) and selectivity (Eq. 11.6) for the catalyst are also verified.

$$C\ (\%) = m_C/m_F \times 100 \tag{11.5}$$

where m_C is the mass of the reactant (g or kg) that reacted and m_F is its initial mass (g or kg).

$$S\ (\%) = m_R/m_C \times 100 \tag{11.6}$$

where m_R is the mass of product formed (g or kg) and m_C is the mass of converted reactant (g or kg).

The application of green chemistry principles should be encouraged through atomic economy, energy saving, less waste generation, and, of course, the use of catalysts instead of stoichiometric reactions. Two parameters that can be applied are atomic economy for the reaction (Eq. 11.7) and the E-Factor to assess possible environmental impact of waste generation (Eq. 11.8) (Lapkin and Constable, 2009):

$$\text{Atomic Economy} = M_P/\Sigma M_R \tag{11.7}$$

where M_P is the molecular mass of the formed product ($g\,mol^{-1}$) and ΣM_R is the sum of the molecular masses of the reactants ($g\,mol^{-1}$).

$$\text{Factor} - \text{E} = m_{res}/m_P \tag{11.8}$$

where m_{res} is the mass of residue formed (g or kg) and m_P is the mass of the target product formed (g or kg).

Homogeneous catalysts for biomass and CO_2 conversion

Table 11.1 presents some homogeneous catalysts for biomass and CO_2 conversion as the RC raw materials.

As observed by Zhao et al. (2021), homogeneous catalysts have a better catalytic performance than that of heterogeneous catalysts due to their uniform dispersion. However, the biggest disadvantage of homogeneous catalysts is that they cannot be effectively recycled after the reaction, which makes it difficult to reduce the cost in industrial applications.

Heterogeneous catalysts for biomass and CO_2 conversion

Fig. 11.2 depicts some commercial heterogeneous catalysts, and Table 11.2 presents some heterogeneous catalysts for biomass and CO_2 conversion.

The conversion of biomass into chemicals was a most explored theme than the CO_2 conversion in recent years, and there are a variety of reactions and heterogeneous catalysts that can be applied. For example, Besson et al. (2014) conducted a review of several heterogeneous catalysts for biomass conversion involving reactions of hydrogenation, isomerization, hydrogenolysis, and oxidation, among others. Additionally, heterogeneous nanocatalysts are promising for biomass valorization; Sudarsanam et al. (2019) stated that porosity and nanostructure are the most versatile features of heterogeneous solid catalysts, e.g., zeolites, which can greatly determine the accessibility of specific active sites, reaction mechanisms, and the selectivity of desirable products.

Furthermore, to develop the best heterogeneous catalysts, computational chemistry using modeling as ab initio can prove the trends and propose novel catalyst design for a variety of cellulosic-based chemicals such as levulinic acid or furanic molecules (Réocreux and Michel, 2018). Moreover, this approach can be applied to catalysts for CO_2 conversion.

Analytical techniques

Monitoring the formation of products and by-products and the consumption of precursor molecules of chemical reactions can be done using gas chromatography (GC) or high-performance liquid chromatography (HPLC), coupled with mass spectrometry with compounds separated according to their polarity, vapor pressure, and solubility. They can then be detected by the spectrometer based on the mass/charge ratio (m/z) of the species formed after processes of molecule fragmentation and ions.

Table 11.1 Status of homogeneous catalysts for biomass and CO_2 conversions.

Catalyst	Substrate	Products	Status	Reference
Soluble $NaVO_3$ homogeneous catalyst in carbonate molten salt	CO_2 for electrochemical conversion	Graphite consisted of carbon nanotubes and sheets	Developed in laboratory scale (TRL3)	Chen et al. (2020)
In situ-generated homogeneous catalysts from a copper-containing coordination polymer, i.e., $((CuL)_n$, L = (Z)-2-(5-chlorin-2-hydroxy benzylideneamino) acetic acid	CO_2 for direct air capture and conversion	Cyclic carbonates	Developed in laboratory scale (TRL3)	Zou et al. (2015)
COOH-functionalized catalyst with mesityl *N*-substituents and a C_3-carbon bridge	CO_2 conversion with epoxydes	Cyclic carbonates	Under development (TRL1–2)	Li et al. (2009)
Diluted aqueous acid solution (H_2SO_4)	Ketoses and aldoses by mean dehydration	5-Hydroxymethylfurfural	Well established (TRL9); however, improvements are welcome	van Putten et al. (2013)

Table 11.2 Status of heterogeneous catalysts for biomass and CO_2 conversions.

Catalyst	Substrate	Product(s)	Status	Reference
Raney-Ni	HMF	DHMTHF by hydrogenation	Developed in laboratory scale (TRL3)	Connolly et al. (2010)
1.6 wt% Ru/C	Glucose	Hydrogenated glucose by H_2 addition	Developed in laboratory scale (TRL3)	Gallezot et al. (1998)
Pt/Al_2O_3	Phenols derived from lignin	Several products by means hydrogenolysis and transalkylation	Developed in laboratory scale (TRL3)	Runnebaum et al. (2012)
H-ZSM-5	Alkaline lignin	Aromatic products by means catalytic fast pyrolysis	Developed in laboratory scale (TRL3)	Liang et al. (2021)
Zeolite-supported single-atom catalysts (SACs) using Sc, Y, La, Ru, Rh, Ni, Pd, and Pt	CO_2	CO, fatty acids, and aldehydes by redox or associative mechanisms	Under development (TRL1–2)	Alonso et al. (2021)
$Cu/ZnO/Al_2O_3$	CO_2	Methanol by hydrogenation with recyclable alcohol-assisted system	Developed in laboratory scale (TRL3)	Sen et al. (2021)
Rosin-based porous heterogeneous catalyst (biomass-derived hyper-cross-linked porous heterogeneous catalysts MPAc-Br and MPAc-OH-Br)	CO_2	Cyclic carbonates by cycloaddition	Developed in laboratory scale (TRL3)	Lai et al. (2021)

HMF, 5-(hydroxymethyl)furfural; *DHMTHF*, 2,5-bis(hydroxymethyl)tetrahydrofuran; *H-ZSM-5*, acid zeolite Socony Mobil (5 angström of pore size); *MPAc-Br*, 3-[4-(Bromomethyl)phenyl]-7-(diethylamino)coumarin.

A

B

Fig. 11.2 Examples of commercial heterogeneous catalysts that can be used to convert RC raw materials into chemicals. (A) Zeolite, a microporous aluminosilicate commonly used as commercial adsorbents and catalysts in several industrial processes (source: internet, *author not identified*); (B) Raney-Nickel catalyst, a solid microcrystalline porous Ni used in chemical processes for hydrogenation or hydrogen reduction of organic compounds. *(Source: https://www.shutterstock.com/pt/image-photo/renea-nickel-known-skeletal-solid-microcrystalline-1981628147.)*

Low molecular weight molecules can be analyzed by GC, while other molecules (semivolatile and nonvolatile) can be analyzed by HPLC.

Spectroscopic identification of the products is carried out using Fourier Transform midinfrared absorption (wave number range from 4000 to 400 cm^{-1}, in KBr pellet), ultraviolet-visible absorption spectrophotometry (wavelength range from 190 at 800 nm), nuclear magnetic resonance of ^{1}H and ^{13}C nuclei (liquid and solid states), and mass spectrometry (Silverstein et al., 2014). Groups and chemical environments are characterized after the absorption of the respective incident radiation producing a characteristic signal to be captured by the detector.

Separations, purifications, and additional physicochemical tests (melting and boiling points, determination of molecular formula, solubility, purity, and refractive index) can be applied if there is a need for a better understanding of the molecular structure (Shriner et al., 1998) by means of the use of classical organic analysis methods, which are based on a measure of the absolute value of a given property.

Kinetic measurements of chemical reactions in an aqueous media, which is usually the environment where reactions with biomass components take place, can be performed using HPLC with a UV-visible absorption detector (Mikkola et al., 1999).

In determining the conversion and selectivity of catalysts, the formation of target products and by-products is also monitored by HPLC with a UV-visible absorption detector (Cybulski and Moulijn, 1998). A characteristic signal associated with consumption is attributed to the reactant (conversion) or to the formation of the product/by-product (selectivity).

The superficial and structural analysis of inorganic catalysts is done using X-ray diffractometry, X-ray photoelectron spectroscopy, or scanning electron microscopy (Richards, 2006). Samples must be solid (powder) without special preparation. The analysis of redox processes of transition metals used as inorganic catalysts applies cyclic voltammetry or polarography (Zoski, 2007).

11.2 Homogeneous catalysis applied to agroindustrial biomass: A case study

This case study is based on well-established biodiesel production by means of transesterification of vegetable oils, e.g., soybean oil. It involves four steps described as follows:

- Biodiesel production begins with the pretreatment of the raw material, that is, the oils are filtered, whether food or pure. In both cases, filtration removes solid contamination and eliminates as much water as possible to avoid hydrolysis, a process responsible for producing a glycerol and three fatty acid molecules.
- In the second stage, the determination and treatment of free fatty acids is carried out, that is, the pure oil sample becomes the standardized base of the mixture, and the residual oil is analyzed to determine the amount of fatty acids present in the sample. The acids found are esterified, and the glycerides are removed through neutralization.
- In the third step, the reactions are carried out, and the standardized base is added slowly to an excess of methanol, or ethanol, and stirred until dissolved. Alcohol is added in excess to increase the yield of esters, thus allowing the formation of the process phase that separates biodiesel from glycerol. For the reaction to take place, a catalyst is used; the most common is a mixture with 0.5% caustic soda in relation to the weight of the oil. Basic catalysts, like the one mentioned earlier, accelerate the reaction approximately 4000 times and are more cost-effective than acid catalysts.
- In the fourth and last step, the purification of the biodiesel takes place, that is, the products generated during its manufacture are removed. Among them are glycerin, wastewater, and methyl esters. To remove impurities from the biofuel, an alternative is washing with hot water. When the catalyst used is basic, washing with acidified water neutralizes it. The liquid compounds can be separated from the esters by decantation and then heated to dry and remove moisture.

The transesterification reaction is described in Scheme 11.1.

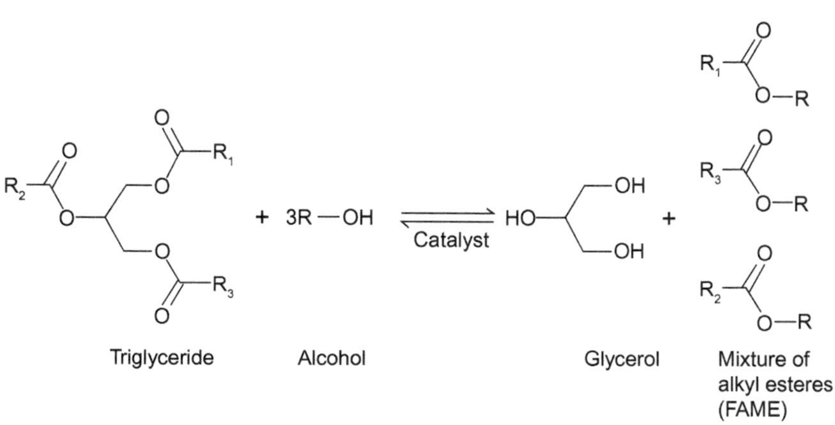

Scheme 11.1 The transesterification reaction. *Credit: author.*

11.3 Homogeneous catalysis applied to CO_2: A case study

This case study is based on the work of Zou et al. (2015) regarding the conversion of CO_2 at atmospheric pressure to cyclic carbonates with in situ-generated homogeneous catalysts from a copper-containing coordination polymer.

A copper-containing coordination polymer ($(CuL)_n$, L=(Z)-2-(5-chlorin-2-hydroxy benzylideneamino) acetic acid, namely BIT-C), was synthesized and structurally analyzed by single-crystal X-ray diffraction. BIT-C presented an outstanding catalytic activity for the conversion of atmospheric pressure CO_2 and various epoxides to cyclic carbonates.

Several quaternary ammonium halides, ionic liquids, and organic bases were considered to be effective co-catalysts. Using BIT-C, Bu_4NBr, and phenyl glycidyl ether, direct capture and conversion of CO_2 from the air was shown to be feasible at room temperature with 19% yield of phenoxy propylene carbonate.

During reaction, multistage dissociation of BIT-C was detected. The binuclear Cu fragment ($(CuL)_2$) was demonstrated to be an essential active species in catalysis and unable to be generated from the raw materials of BIT-C but currently only from the multistage dissociation of BIT-C.

Kinetic analysis and density functional theory (DFT) calculations suggested a reaction mechanism where a copper center activates epoxides and CO_2 successively.

11.4 Heterogeneous catalysis applied to agroindustrial biomass: A case study

This case study is based on the work of Vaz Jr. et al. (2017) regarding selective hydrogenation of D-xylose from sugarcane bagasse to xylitol using palladium catalyst supported on carbon nanostructured composite.

A method was developed of preparing palladium impregnated carbon nanostructured materials or carbon nanotubes (Pd/CNT) and evaluating their catalytic activity in the D-xylose hydrogenation reaction into xylitol. It was found that a 4.9% Pd/CNT catalyst, prepared by this method, showed good catalytic activity for the selective hydrogenation of D-xylose, converting most of the substrate (80%) after 3 h of reaction at 150°C into xylitol with good selectivity (85%). These results indicate a promising future for this type of catalyst, which may in the future replace the catalysts used industrially to carry out the hydrogenation of D-xylose to xylitol. Furthermore, the values obtained with this catalyst were higher than the values of the commercial catalyst 5% Pd/C.

The difference in selectivity between Pd/C and Pd/CNT occurs because of several factors:

- The existence of a peculiar metal-support interaction between the Pd nanoparticles and the surface of the CNT, that is, the electronic modification that occurs through the transfer of electrons between the metal and support modify the adsorption and selectivity of the products.
- The concentration of oxygenated groups on the catalytic supports—which is low on the surface of the CNT and high on the surface of the carbon—can significantly affect the catalyst's catalytic selectivity by modifying the adsorption mode of the reactant.
- The large amount of microporosity in the carbon support (which is absent in the CNT) can induce readsorption and cause consecutive undesirable reactions. Thus, the presence of micropores in the catalytic support can artificially increase the residence time of reactants and products and, consequently, modify their adsorption properties and alter the natural path of the hydrogenation reaction.

11.5 Heterogeneous catalysis applied to CO_2: A case study

This case study is based on the work of Wang et al. (2018) regarding conversion of CO_2 to lower hydrocarbons (C_2-C_4) over bifunctional catalysts composed of In_2O_3-ZrO_2 and zeolite.

Although many researchers have reported CO_2 hydrogenation to various C_1 chemicals, it is still challenging to convert CO_2 directly and selectively to C_2-C_4 hydrocarbons in terms of overcoming the extreme inertness of CO_2 and a high C-C coupling barrier.

This work reported an efficient integration of methanol-synthesis and the methanol-to-hydrocarbons with the bifunctional catalyst component of In_2O_3-ZrO_2 and SAPO-5 (silicoaluminium phosphates). These tandem reactions exhibit an excellent relative selectivity of C_2-C_4 (83%) with a suppressed CH_4 relative selectivity less than 3% at 300°C. A detailed analysis indicates that the partially reduced indium oxide surface (In_2O_3-ZrO_2) can better activate CO_2 and promote the synthesis of methanol than In_2O_3 alone, and C-C coupling is subsequently manipulated within the confined acidic pores of SAPO-5 according to XRD, H_2-TPR, CO_2-TPD, SEM, and TEM analyses.

Furthermore, the proximity of two components and the content of Si also play an important role in such outstanding selectivity to C_2-C_4. This study can pave a new path for the direct synthesis of lower hydrocarbons.

11.6 Conclusions

The development and application of homogeneous and heterogeneous catalysts are pivotal for the exploration of the biomass and CO_2 potential as RC raw materials for industrial purposes.

However, the majority of the observed catalysts are under research and development in reduced values for the TRL scale, requiring more efforts to reach industrial scale.

References

Alonso, G., López, E., Huarte-Larrñaga, F., Sayós, R., Prats, H., Gamallo, P., 2021. Zeolite-encapsulated single-atom catalysts for efficient CO_2 conversion. J. CO_2 Util. 54, 101777.

Besson, M., Gallezot, P., Pinel, C., 2014. Conversion of biomass into chemicals over metal catalysts. Chem. Rev. 114, 1827–1870.

Chen, Y., Wang, M., Lu, S., Tu, J., Jiao., 2020. Electrochemical graphitization conversion of CO_2 through soluble $NaVO_3$ homogeneous catalyst in carbonate molten salt. Electrochim. Acta 331, 135461.

Connolly, T.J., Considine, J.L., Ding, Z., Forsatz, B., Jennings, M.N., MacEwan, M.F., McCoy, K.M., Place, D.W., Sharma, A., Sutherland, K., 2010. Efficient synthesis of 8-oxa-3-azabicyclo[3.2.1]octane hydrochloride. Org. Process. Res. Dev. 14, 459–465.

Cybulski, A., Moulijn, J.A. (Eds.), 1998. Structured Catalysts and Reactors. Marcel Dekker, New York. 645 pp.

Gallezot, P., Nicolaus, N., Fleche, G., Fuertes, P., Perrard, A., 1998. Glucose hydrogenation on ruthenium catalysts in a trickle-bed reactor. J. Catal. 180, 51–55.

International Union of Pure and Applied Chemistry, 2021a. The IUPAC compendium of chemical terminology. Catalyst. https://goldbook.iupac.org/terms/view/C00876. December 2021.

International Union of Pure and Applied Chemistry, 2021b. The IUPAC Compendium of Chemical Terminology. Rate of Reaction. https://goldbook.iupac.org/terms/view/R05156. December 2021.

Lai, S., Gao, J., Xiong, X., 2021. Rosin-based porous heterogeneous catalyst functionalized with hydroxyl groups and triazole groups for CO_2 chemical conversion under atmospheric pressure condition. React. Funct. Polym. 165, 104976.

Lapkin, A., Constable, D. (Eds.), 2009. Green Chemistry Metrics – Measuring and Monitoring Sustainable Processes. Blackwell, Chichester, pp. 69–199.

Li, Y., Dominelli, B., Reich, R.M., Liu, B., Kühn, F.E., 2009. Bridge-functionalized bisimidazolium bromides as catalysts for the conversion of epoxides to cyclic carbonates with CO_2. Catal. Commun. 124, 118–122.

Liang, J., Shan, G., Sun, Y., 2021. Catalytic fast pyrolysis of lignocellulosic biomass: critical role of zeolite catalysts. Renew. Sust. Energ. Rev. 139, 110707.

Mikkola, J.-P., Sjöholm, R., Salmi, T., Mäki-Arvela, P., 1999. Xylose hydrogenation: kinetic and NMR studies of the reaction mechanisms. Catal. Today 48, 73–81.
Réocreux, R., Michel, C., 2018. Rational design of heterogeneous catalysts for biomass conversion – inputs from computational chemistry. Curr. Opin. Green Sustain. Chem. 10, 51–59.
Richards, R. (Ed.), 2006. Surface and Nanomolecular Catalysis. Taylor & Francis, Boca Ranton.
Runnebaum, R.C., Nimmanwudipong, T., Block, D., Gates, B.C., 2012. Catalytic conversion of compounds representative of lignin-derived bio-oils: a reaction network for guaiacol, anisole, 4-methylanisole, and cyclohexanone conversion catalysed by Pt/γ-Al_2O_3. Catal. Sci. Technol. 2, 113–118.
Sen, R., Koch, C.J., Galvan, V., Entesari, N., Goeppert, A., Prakash, G.K.S., 2021. Glycol assisted efficient conversion of CO_2 captured from air to methanol with a heterogeneous $Cu/ZnO/Al_2O_3$ catalyst. J. CO_2 Util. 54, 101762.
Shriner, R.L., Fuson, R.C., Curtin, D.Y., Morril, T.C., 1998. The Systematic Identification of Organic Compounds, seventh ed. Wiley, New York.
Silverstein, R.M., Webster, F.X., Kiemle, D.J., Bryce, D.L., 2014. Spectrometric Identification of Organic Compounds, eighth ed. Wiley, New York.
Sudarsanam, P., Peeters, E., Makshina, E.V., Parvulescu, V., Sels, B.F., 2019. Advances in porous and nanoscale catalysts for viable biomass conversion. Chem. Soc. Rev. 48, 2366–2421.
van Putten, R.-J., Soetedjo, J.N.M., Pidko, E.A., van der Waal, J.C., Hensen, E.J.M.H., de Jong, E., Heeres, H.J., 2013. Dehydration of different ketoses and aldoses to 5-hydroxymethylfurfural. ChemSusChem 6, 1681–1687.
Vaz Jr., S., Donate, P.M., da Silva, W.R., Kikuchi, D.K., de Souza, V.S., 2017. Hidrogenação Seletiva de D-xilose a Xilitol Utilizando Catalisadores de Paládio Suportados em Compósito Nanoestuturado de Carbono (Selective Hydrogenation of D-xylose to Xylitol Using Palladium Catalysts Supported on Carbon Nanostructured Composite). In: Boletim de Pesquisa e Desenvolvimento 04. https://ainfo.cnptia.embrapa.br/digital/bitstream/item/161170/1/BPD-04-CNPAE.pdf. (Accessed December 2021. in Portuguese.
Wang, J., Zhang, A., Jiang, X., Song, C., Guo, X., 2018. Highly selective conversion of CO_2 to lower hydrocarbons (C_2-C_4) over bifunctional catalysts composed of In_2O_3-ZrO_2 and zeolite. J. CO_2 Util. 27, 81–88.
Zhao, Y., Lu, K., Xu, H., Zhu, L., Wang, S., 2021. A critical review of recent advances in the production of furfural and 5-hydroxymethylfurfural from lignocellulosic biomass through homogeneous catalytic hydrothermal conversion. Energy Rev. 139, 110706.
Zoski, C.G., 2007. Handbook of Electrochemistry. Elsevier, Amsterdam.
Zou, B., Hao, L., Fan, L.-Y., Gao, Z.-M., Chen, S.-L., Li, H., Hu, C.-W., 2015. Highly efficient conversion of CO_2 at atmospheric pressure to cyclic carbonates with *in situ*-generated homogeneous catalysts from a copper-containing coordination polymer. J. Catal. 329, 119–129.

CHAPTER 12

Remarks and conclusions of the technological aspects for renewable carbon

As commented in Chapter 5 (Part 1), remarks and conclusions are approaches used to facilitate book understanding and its applicability in real situations. Remarks cover the most representative statements described in the text while conclusions critically summarize the conducted discussions. Then both can be used as a practical compilation, which take parts of relevant contents from each chapter. This didactic approach comprises:

- The reading of the text
- The understanding of its content
- The application of its fundamentals to the renewable carbon (RC) concept related to technological basis offered by Part 2

12.1 Remarks

We saw in Chapter 6 that the sources of RC materials can be related to the following industrial sectors: (i) agriculture and agroindustry for biomass, their products and residues; (ii) energy generation from fossil fuels, chemical industry, iron, steel, and cement for CO_2; and (iii) plastics (in a general way) comprised of resins and synthetic fibers for recyclable materials. The main technologies for conversion of RC raw materials into valuable products are: immobilized enzymes; synthetic biology; batch synthesis, with or without catalysts; flow synthesis, with or without catalysts; combustion; gasification; and pyrolysis. Regarding challenges to establish technologies for RC raw materials conversion, they could rise from: difficulty to establish RC sources as competitive industrial feedstocks as alternative to oil, which reduces the opportunity for conversion technologies related to these sources; high costs of industrial plants dedicated to RC, which tends to be reduced with increased usage; scarce commercial availability of dedicated

Renewable Carbon
https://doi.org/10.1016/B978-0-323-99735-5.00011-6

catalysts for biomass and CO_2 conversion, mainly for chemical processes; and absence of governmental support for renewable or green products development and commercialization. However, these same challenges can turn into opportunities for the RC approach because they can develop new technologies and value chains and spread wealth in society.

We saw in Chapter 7 that carbon dioxide capture can be carried out in three stages (Intergovernmental Panel on Climate Change, 2005): precombustion; combustion; and postcombustion. The main classes of technologies for carbon capture and usage (CCU) and carbon capture, usage, and storage (CCUS) are related to adsorption technologies, absorption technologies, membrane separation technologies, and chemical capture technologies. For these technologies, assessment by means the technology readiness level (TRL) is very useful and considered to clarify that all elements of the carbon capture and storage value chain are mature and have been in commercial operation for decades. However, incremental improvements in those technologies have—and will continue—to reduce their cost. Furthermore, net-zero energy system by 2050 should ensure stable and affordable energy supplies, providing universal energy access, and enabling robust economic growth (International Energy Agency, 2021). To reach these goals, CCU/CCUS technologies certainly have a relevant role.

We saw in Chapter 8 that nanoscience (the branch definition) and nanotechnology (the branch application) are subjects of interest from academia and industry due to the potential for the development of new materials to generate new value chains such as those derived from RC raw materials. For example, the development of catalysts as functionalized nanoparticles (NPs) obtained from biomass is extremely strategic to convert the biomass into valuable products and to meet green chemistry principles (Shuttleworth et al., 2014). Particle dimensions in the range of 1–100 nm (International Standard Organization, 2008) can confer some interesting molecular properties to be explored in several applications such as (Hassellöv and Kaegi, 2009): surface area; porosity structure; crystallinity; size distribution; surface charge; and surface speciation. It is expected a TRL range comprised from TRL3 (development) to TRL6 (demonstration) is appropriate for most cases observed in the scientific and patent literature and industry statements.

We saw in Chapter 9 that the bioeconomy model demands a set of technologies to convert biomass into valuable products. From this scenario, industrial biotechnology (i.e., enzymatic and microbial routes) is expected to achieve prominence when compared to other technology classes for some products, such as organic acids and alcohols. Generally, the industrial application for RC raw materials conversion comprises fermentation technologies for the conversion step, e.g., yeasts; biocatalysis technologies also for the conversion step, e.g., immobilized enzymes; and separation technologies for the downstream step, e.g., activated carbon, centrifugation, and chromatography. Some processes described in this chapter are already well established (TRL9). However, for some compounds of great industrial interest, i.e., the use of biochemical processes, is still in the proof-of-concept phase (TRL3). On the other hand, some compounds are being abandoned.

We saw in Chapter 10 that thermochemistry for industrial purposes applied to RC raw materials comprises a set of technologies and processes based on different temperature ranges, e.g., combustion, slow pyrolysis, torrefaction, fast pyrolysis, flash pyrolysis, and gasification. It can be very useful to treat biomass and plastic residues to reduce environmental impacts and develop new products. We can observe that the energy cost can be the main parameter to be considered for the operational expenditure (OPEX) in an economic feasibility assessment. Energy is a very expensive input for the industry according to its origin: hydroelectricity is cheaper than thermoelectricity because the latter uses fossil fuels (e.g., coal and natural gas) when the former uses surface water. Sustainable chemistry, the chemistry developed and executed according the sustainability components (discussed in Part 3) can be answered by means the use of thermochemical conversion to obtain chemicals, materials, and energy of the second and third generations considering the carbon footprint.[a]

And we saw in Chapter 11 that catalysis can be classified as homogeneous and heterogeneous catalysis; in the first, only one phase is involved, and in the second, the reaction occurs at or near an interface between phases. The conversion of biomass into chemicals was more explored than CO_2 conversion in recent years, and there are a variety of reactions and heterogeneous catalysts to be applied for the former. Additionally, heterogeneous nanocatalysts are promising for biomass valorization; Sudarsanam et al. (2019) stated that porosity and nanostructure are the most versatile features of heterogeneous solid catalysts, e.g., zeolites, which can greatly determine the accessibility of specific active sites, reaction mechanisms, and the selectivity of desirable products. Furthermore, to develop the best heterogeneous catalysts, computational chemistry using modeling as ab initio can prove the trends and propose novel catalysts design (Réocreux and Michel, 2018). Case studies then demonstrated that heterogeneous and

[a] https://unesdoc.unesco.org/ark:/48223/pf0000217181.

homogeneous catalysts can be applied to biomass and CO_2 conversion into chemicals.

Thus, these six chapters depicted a wide scope of conversion technologies applicable to RC raw materials.

12.2 Conclusions

We concluded in Chapter 6 that RC sources comprise huge value products and value chains, as depicted by the sectors of agriculture and agroindustry, energy generation, chemical industry, iron, steel, cement, plastics (in a general way), resins, and synthetic fibers. The use of RC raw materials (CO_2, biomass, and recyclable materials) depends on the application of technologies related to biochemical, organic, inorganic, and thermochemical synthetic routes, which can be evaluated by means of the TRL assessment approach. However, there are observed challenges related to science and economy, among other aspects, to be considered for technological advance. Finally, the advantages of RC as industrial feedstock overcomes its disadvantages for sustainable development.

We concluded in Chapter 7 that technologies for CCU/CCUS are paramount for industrial and energy sectors to permit CO_2 mitigation and its use as RC raw material. Despite its importance, CCU/CCUS approaches comprise technology categories with limitations and challenges for their application. However, we can observe an increase in their use and the development of more sustainable and affordable technologies.

We concluded in Chapter 8 that nanotechnology is a development and application strategy that can promote a valorization of CO_2, biomass components, and recycling materials, despite its unclear risks to the environment and public health. Products such as catalysts, absorbents, nanotubes, and graphene can be obtained for application in several sectors such as medicine, consumer products, energy, materials, and manufacturing. Furthermore, the application of nanotechnology to the recycling of plastics can expand its technological and social potentials.

We concluded in Chapter 9 that biotechnological technologies and processes can be used for RC conversion into valuable products. Some technologies are under development, and others are available for the industry. However, limitations are observed related to purity and yields of the final products, which should be overcome to reach economic feasibility when considering operating and capital costs (CAPEX and OPEX).

We concluded in Chapter 10 that thermochemical conversion, despite the high demand of energy to make it an ongoing process, can be seen as a smart set of technologies to treat and reuse biomass and plastic wastes. For example, it is expected to reach 100% carbon recovery. Then RC can be explored to produce chemicals, materials, and energy of second and third generation obeying sustainable chemistry.

Finally, we concluded in Chapter 11 that the development and application of homogeneous and heterogeneous catalysts are pivotal for the exploration of biomass and CO_2 potential as RC raw materials for industrial purposes. However, a majority of the observed catalysts are under research and development in reduced values for the TRL scale, requiring more efforts to reach an industrial scale.

Thus, technologies to convert RC raw materials into more valuable products are not completely established, but they have high potential to contribute to a more sustainable economy and society.

References

Hassellöv, M., Kaegi, R., 2009. Analysis and characterization of manufactured nanoparticles in aquatic environments. In: Lead, J.R., Smith, E. (Eds.), Environmental and Human Impacts of Nanotechnology. Willey, Chichester.

Intergovernmental Panel on Climate Change, 2005. IPCC Special Report on Carbon Dioxide Capture and Storage. Prepared by Working Group III of the Intergovernmental Panel on Climate Change. Cambridge University Press, Cambridge. 442 pp.

International Energy Agency, 2021. Net Zero by 2050: A Roadmap for the Global Energy Sector. https://www.iea.org/reports/net-zero-by-2050. (Accessed December 2021).

International Standard Organization, 2008. ISO/TS 27687:2008: Nanotechnologies: Terminology and Definitions for Nano-objects: Nanoparticle, Nanofiber and Nanoplate. ISO, Geneva.

Réocreux, R., Michel, C., 2018. Rational design of heterogeneous catalysts for biomass conversion – inputs from computational chemistry. Curr. Opin. Green Sustain. Chem. 10, 51–59.

Shuttleworth, P.S., De Bruyn, M., Parker, H.L., Hunt, A.J., Budarin, V.L., Matharu, A.S., Clark, J.H., 2014. Applications of nanoparticles in biomass conversion to chemicals and fuels. Green Chem. 16, 573–584.

Sudarsanam, P., Peeters, E., Makshina, E.V., Parvulescu, V., Sels, B.F., 2019. Advances in porous and nanoscale catalysts for viable biomass conversion. Chem. Soc. Rev. 48, 2366–2421.

PART 3

Sustainability

CHAPTER 13

Introduction to the sustainability concept applied to renewable carbon

The sustainability of products and processes is a condition pursued in the 21st century to reduce or eliminate negative impacts on the environment and public health from the industrial sectors, especially those related to carbonaceous substances (chemicals, materials, fuels, etc.).

However, scientific, technological, and economic aspects should work side-by-side to reach this condition, as considered in this chapter.

13.1 The sustainability concept and its components

Sustainability is a complex concept that needs to be addressed by the global society in the 21st century. In our technological world, there is a continuous search for the development of sustainable processes and products that can produce more valuable goods. Probably the most suitable definition comes from the UN World Commission on Environment and Development: "*Sustainable development is development that meets the needs of the present without compromising the ability of future generations to meet their own needs.*" That is, sustainability looks to future resources and life quality by means of smart strategies, which involves the 17 Sustainable Development Goals (SDGs) (United Nations, 2021). These SDGs are depicted in Fig. 13.1.

Goals 6 (clean water and sanitation), 7 (affordable and clean energy), 9 (industry, innovation, and infrastructure), 12 (responsible consumption and production), and 13 (climate action) are closely related to renewable carbon (RC) sources and processing to achieve sustainable productive chains.

For didactic purposes, the sustainability concept is comprised of three components or pillars:

- Economic impacts
- Social impacts
- Environmental impacts

Renewable Carbon
https://doi.org/10.1016/B978-0-323-99735-5.00005-0

Fig. 13.1 The 17 Sustainable Development Goals (SDGs) (United Nations, 2021). *(Reprinted with permission from the United Nations.)*

These impacts can be *positive* or *negative* according to the appropriate metrics and are considered in detail in Part 3.

The three pillars or components of sustainability are composed of a variety of internal indicators to be considered, such as:

- Economic impacts: natural resources use, environmental management, and pollution prevention applied to air, water, land, and waste
- Social impacts: standard of living, education, community, and equal opportunities
- Economic impacts: profit, cost savings, economic growth, and research and development

Several indicators can be listed according to the productive chain, making them suitable for each case of study, i.e., they should meet the demand. Furthermore, Sikdar (2003) considered the intersection—or interdependencies—of the pillars with more internal indicators, which can generate:

- Environmental-economic: energy efficiency, subsides/incentives for use of natural resources
- Social-environmental: environmental laws, natural resources stewardship
- Economic-social: business ethics, fair trade, worker's rights

Then a positive impact is expected from all three sustainability components to deliver more environmentally friendly and renewable consumer goods to society.

When we consider the processing of RC raw materials, there are some categories of specific sustainability indicators available for use, mainly for the environmental and social components to achieve a reduced environmental footprint and increased societal value (United States Environmental Protection Agency, 2012).

For a reduced environmental footprint (United States Environmental Protection Agency, 2012):

- Greenhouse gas emissions (GHG): reduction or absence
- Persistent toxic emissions, such as dioxins and furans: reduction or absence
- Material intensity: reduction
- Ecological impacts: reduction or absence
- Land use: reduction
- Water intensity: reduction or absence
- Energy intensity: reduction

Regarding GHG emissions, it is related mainly to carbon dioxide (CO_2) and methane (CH_4) released into the atmosphere. Persistent toxic emissions, for example, include the release of dioxins and furans (gaseous effluent). For material intensity, the amount of feedstock used in the processing must be considered, as well as factors such as carbon dioxide released by tillage during land use.

For increased societal value (United States Environmental Protection Agency, 2012):

- Poverty alleviation: improvement
- Health and safety improving: improvement
- Asset recovery: improvement
- Prosperity and economic resilience: improvement
- Biodiversity and ecological resilience: improvement
- Resource conservation: improvement
- Human dignity: improvement

As observed by Vaz Jr. (2020), to meet the goals for reduced environmental footprint and increased societal value, it is paramount to invest in research, development, and innovation (R&D&I) to create new "green" technologies, especially for production systems and transformation processes.

These new technologies should involve the reduction in water, energy, and inputs use allied to cutting-edge scientific assets from areas such as nanotechnology and biotechnology.

13.2 Circular economy approach for renewable carbon

As previously introduced in Chapter 1, the concept of circular economy and its application can contribute to a holistic vision of the economic context of harnessing industrial and agroindustrial residues. The European Parliament (2021) defines it as, "*a model of production and consumption, which involves sharing, leasing, reusing, repairing, refurbishing and recycling existing materials and products as long as possible. In this way, the lifecycle of products is extended. In practice, it implies reducing waste to a minimum. When a product reaches the end of its life, its materials are kept within the economy wherever possible. These can be productively used again and again, thereby creating further value.*"

Considering these aspects, the circular economy approach can be explored for the RC valorization by different means, such as:

- The creation of new value chains for all RC sources (biomass, CO_2, and recyclable materials)
- The reuse and recyclability for recyclable materials as plastics
- The treatment and mitigation of pollution for biomass, plastic wastes, and CO_2

13.3 Green chemistry approach for renewable carbon usages

As previously introduced also in the Chapter 1, the 12 fundamental principles of green chemistry (GC) are as follows (ACS Green Chemistry Institute, 2021):

(I) Prevention: it is better to prevent waste than to treat or clean up waste after it has been created.

(II) Atom economy: synthetic methods should be designed to maximize the incorporation of all materials used in the process into the final product.

(III) Less hazardous chemical syntheses: wherever practicable, synthetic methods should be designed to use and generate substances that possess little or no toxicity to human health and the environment.

(IV) Designing safer chemicals: chemical products should be designed to affect their desired function while minimizing their toxicity.

(V) Safer solvents and auxiliaries: the use of auxiliary substances (e.g., solvents, separation agents, etc.) should be made unnecessary wherever possible and innocuous when used.

(VI) Design for energy efficiency: energy requirements of chemical processes should be recognized for their environmental and economic impacts and should be minimized. If possible, synthetic methods should be conducted at ambient temperature and pressure.

(VII) Use of renewable feedstocks: a raw material or feedstock should be renewable rather than depleting whenever technically and economically practicable.

(VIII) Reduce derivatives: unnecessary derivatization (use of blocking groups, protection/deprotection, and temporary modification of physical/chemical processes) should be minimized or avoided if possible, because such steps require additional reagents and can generate waste.

(IX) Catalysis: catalytic reagents (as selective as possible) are superior to stoichiometric reagents.

(X) Design for degradation: chemical products should be designed so that, at the end of their function, they break down into innocuous degradation products and do not persist in the environment.

(XI) Real-time analysis for pollution prevention: analytical methodologies need to be further developed to allow for real-time, in-process monitoring and control prior to the formation of hazardous substances.

(XII) Inherently safer chemistry for accident prevention: substances and the form of a substance used in a chemical process should be chosen to minimize the potential for chemical accidents, including releases, explosions, and fires.

Principles IV (design safer chemicals) and VII (use of renewable feedstocks) are those to be more easily explored in the processing of RC materials. However, principle IX (use of catalysts) is also very important because it is related to the conversion processes for RC, as presented in Chapter 11.

Fig. 13.2 illustrates some hypotheses to be analyzed for the application of GC principles in the processing of RC raw materials, as well as in the use of its derived products.

Thus, for the best application of the 12 principles, an exercise is suggested to establish what products and processes are most adequate to be developed/produced/applied in light of such principles and answering some questions:

- What principles are feasible to apply to the process?
- What principles are feasible to apply to the product?
- Can I improve the performance?
- Can I improve the sustainability?
- How much will it cost?

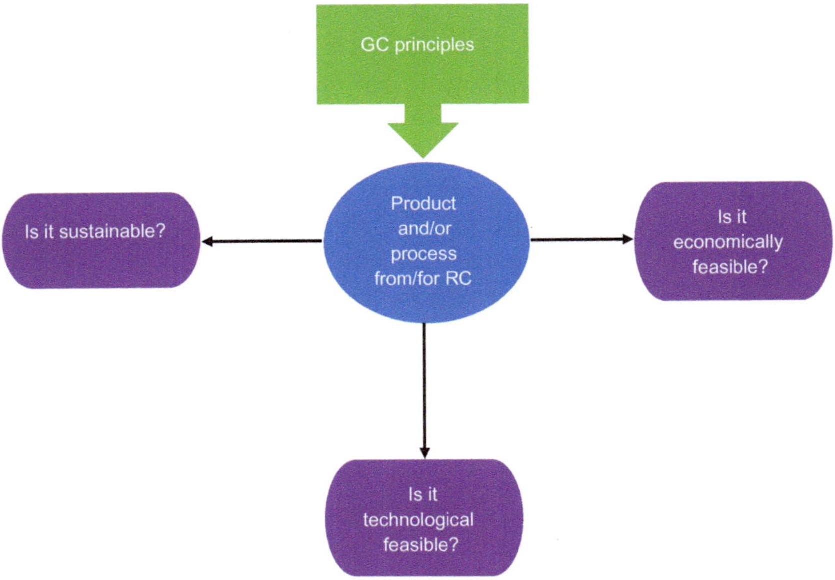

Fig. 13.2 The critical application of green chemistry principles on processes and products related to renewable carbon raw materials. *(Credit: author.)*

Not all results will likely be favorable, but their understanding is an asset to promote the principles in the industry.

A process should be economically and technologically feasible to generate products to be used by the market and to innovate. Furthermore, it is extremely desirable to reach the sustainability for products and processes that will promote environmental gains and a marketing differentiation.

13.4 Anthropogenic and biogenic carbons and their influence on the climate change

Currently, there are two categories of carbon considered for climate change purposes: anthropogenic carbon and biogenic carbon.

Biogenic carbon refers to carbon that is sequestered from the atmosphere during biomass growth, i.e., CO_2, and may be released back into the atmosphere later due to combustion of the biomass or its decomposition (Stamford, 2020). On the other hand, anthropogenic carbon is generated by means of human activities, such as wastewater-derived organic carbon (Golubkov et al., 2019).

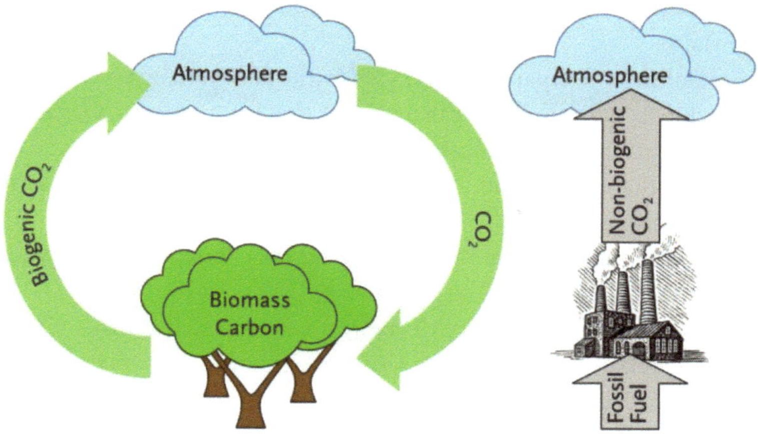

Fig. 13.3 The atmospheric cycles for biogenic carbon *(left)* and nonbiogenic or anthropogenic carbon *(right)*. *(Credit: International Energy Agency—Bioenergy (https://www.ieabioenergy.com/iea-publications/faq/woodybiomass/biogenic-co2/). Reprinted with permission from the International Energy Agency.)*

The renewability of biogenic carbon is visible because it is recycled by nature, while for anthropogenic carbon, it is evaluated and assigned case by case according to its origin and composition (e.g., in principle, fossil fuels are not renewable).

Biogenic carbon can contribute directly to the reduction of climate change through the capture and storage of CO_2 by means of the process of photosynthesis performed by plants for their growth, as depicted in Fig. 13.3.

A direct influence of anthropogenic carbon on climate change is then expected by means of the availability of CO_2 in the atmosphere for the GHG effect, as discussed in Chapter 7.

13.5 The footprint carbon for products and processes

The quantity of emitted and captured GHG, i.e., carbon dioxide and methane, is a factor to be considered both for products and processes as a balance of mass to estimate the sustainability.

According Fullana et al. (2009), carbon footprint (CF) is a method used to quantify the amount of GHG emissions associated with a company (corporate carbon footprint [CCF]) or with the lifecycle of an activity or a product/service (product carbon footprint [PCF]) to determine its contribution to climate change. It is a characteristic explored to reach high percentages of marked share.

To establish and organize the basis for the lifecycle assessment (LCA), which includes the CF in its calculations, the International Organization for Standardization (ISO) published norm 14040:2006 (International Organization for Standardization, 2006). This norm is the global standard for LCA.

Fig. 13.4 depicts the general methodology for the CF calculation based on ISO norm 14040:2006.

Regarding carbon neutrality of the sources of RC (biomass, CO_2, and recyclable materials) obtained by means the CF, some authors suggest a certain prudence to consider it for processing and final products:

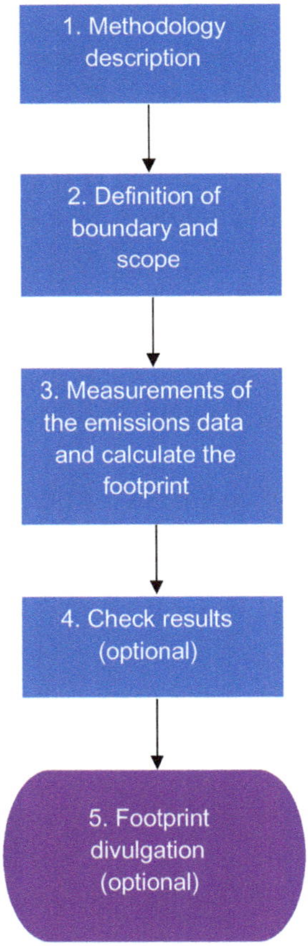

Fig. 13.4 Sequential steps involved in the carbon footprint calculation according to ISO norm 14040:2006. *(Credit: author.)*

- For biomass-derived products: in the case of biofuels, they are not always carbon neutral. Indeed, Johnson (2009) observed that they can, in some cases, be far more carbon positive than fossil fuels.
- For CO_2 from energy sectors consuming fossil fuels: the energy sector contributes about 40% of global emissions of CO_2. Three-quarters of those emissions come from six major economies. Although coal-fired plants account for just 40% of the world's energy production, they were responsible for more than 70% of energy-sector emissions in 2010 (World Bank, 2021a,b).
- For recyclable materials (plastics): Zheng and Suh (2019) calculated that the global lifecycle GHG emissions of conventional plastics were 1.7 Gt of CO_2-equivalent (CO_2eq) in 2015, which would grow to 6.5 GtCO_2eq by 2050 under the current trajectory of their production and use.

From these observations, the need to establish mitigation—the reducing risk of loss of environmental quality from the occurrence of any undesirable event—and reuse strategies is irrefutable for industrial processes and their generated products.

13.6 The carbon market

With the negative impacts rising from climate change, mechanisms for atmospheric pollution control are even more focused on mitigating GHG emissions. Kim and Park (2018) argue that the mechanisms created in world conventions such as the Kyoto Protocol[a] are essentially due to the increase in GHG emissions, especially carbon dioxide. Thus, global concern with the emissions of this gas goes beyond global warming, gaining an economic and social focus. That means, in theory, we can generate sustainable wealth from the residues, i.e., the GHG.

The carbon market comprises agreements made at world conventions by several countries with the objective of promoting and boosting searches for mitigating processes and also to provide a union aimed at raising awareness of GHG reductions (Godoy, 2013).

As a way of encouraging the reduction of GHGs, this market allows carbon transactions between countries that have not reached their reduction quota and those that have exceeded it, with these transactions generating

[a] https://unfccc.int/kyoto_protocol.

revenues for those who have carbon credits (Anand et al., 2019), making them of great economic and social interest.

Understanding the economic viability of the carbon mitigation market will help to estimate, more accurately, the anthropogenic emissions of CO_2 and CH_4 (as discussed in Section 13.4). Over the years, climate change will promote more aggressive effects, with the carbon market being a stimulus for countries to comply with the agreements made in international conventions, in addition to promoting research on the reduction of GHG and their usages (World Health Organization, 2018).

Currently, carbon pricing systems mainly covers industrial, energy, and land use change-related sectors because they are well-established production systems with a certain facility for valuation and certification. Examples of international carbon pricing efforts are:

- The Partnership for Market Readiness (PMR)[b]
- The Carbon Pricing Leadership Coalition (CPLC)[c]
- The Mitigation Action Assessment Protocol (MAAP)[d]

Fig. 13.5 depicts a global map with initiatives to establish an international carbon market.

The establishment of a global market must boost GHG mitigation toward a more sustainable society.

13.7 Conclusions

Ensuring that sustainable development meets the needs of the present without compromising the ability of future generations to meet their own needs by reducing the negative impacts of modern society on the environment is paramount, as highlighted and proposed by the Sustainable Development Goals.

To reach the sustainability condition, strategies such as circular economy, GC, and carbon footprint calculation should be put in practice for the RC sources for industrial usage and valorization. In addition, it can be advantageous not only from an environmental point of view but also from an economic one as observed in carbon pricing efforts.

[b] https://www.thepmr.org/.
[c] https://www.carbonpricingleadership.org/.
[d] https://maap.worldbank.org/#/homepage.

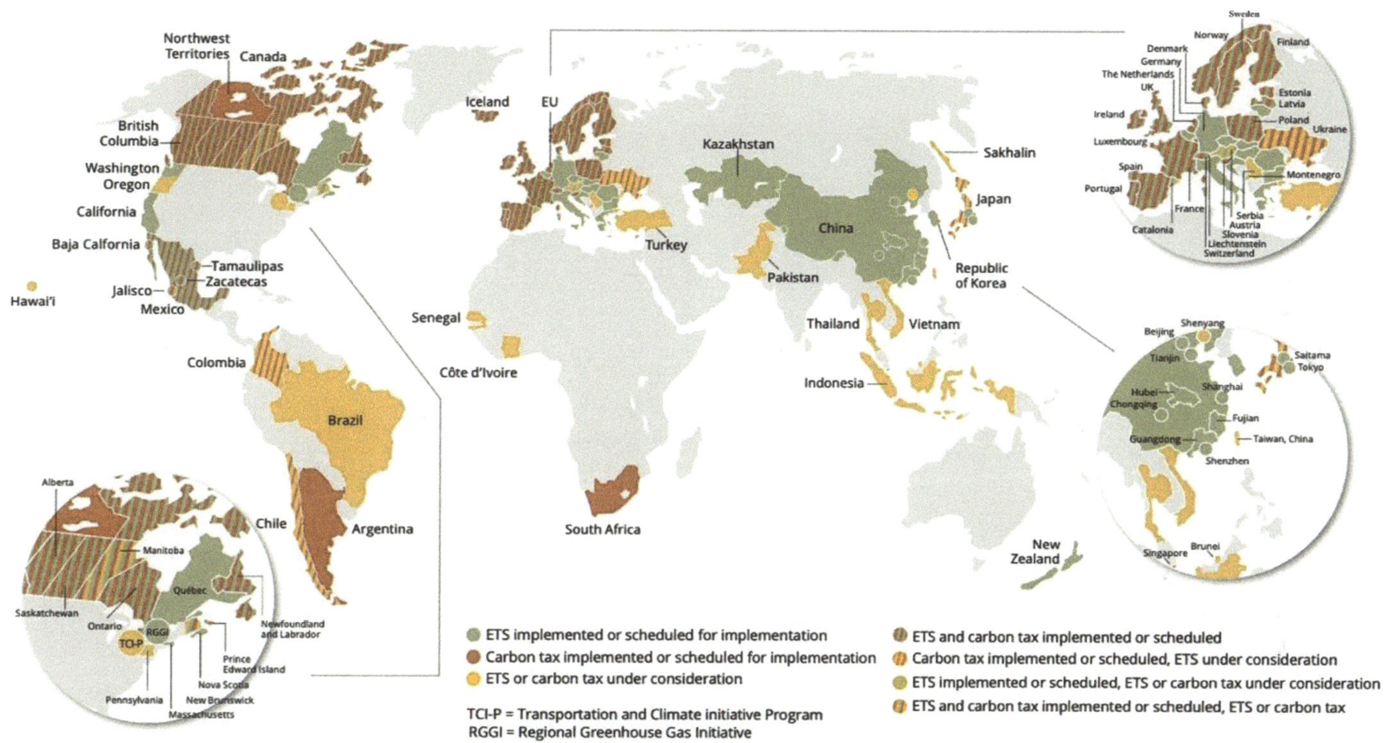

Fig. 13.5 Carbon pricing map for the year of 2021. *ETS*, emission trade system. *(Credit: World Bank, 2021a. State and Trends of Carbon Pricing 2021. Washington, DC, World Bank. World Bank. https://openknowledge.worldbank.org/handle/10986/35620, (Accessed December 2021), World Bank, 2021b Understanding CO_2 Emissions From the Global Energy Sector. https://openknowledge.worldbank.org/handle/10986/17143?show=full, (Accessed December 2021). Reprinted with permission from the World Bank.)*

References

ACS Green Chemistry Institute, 2021. 12 Principles of Green Chemistry. https://www.acs.org/content/acs/en/greenchemistry/principles/12-principles-of-green-chemistry.html. access December 2021.

Anand, N., Duin, R., Tavasszy, L., 2019. Carbon credits and urban freight consolidation: an experiment using agent based simulation. Res. Transp. Econ. 85, 100797.

European Parliament, 2021. Circular Economy. https://www.europarl.europa.eu/thinktank/infographics/circulareconomy/public/index.html. access December 2021.

Fullana, P., Betz, M., Hischier, R., Puig, R., 2009. Life Cycle Assessment Applications: Results from COST Action 530. AENOR, Madrid.

Godoy, S.G.M., 2013. Projetos de redução de emissões de gases de efeito estufa: desempenho e custos de transação (Greenhouse gas emissions reduction projects: performance and transaction costs). Rev. Adm. São Paulo 48, 310–326.

Golubkov, S.M., Golubkov, M., Tiunov, A.V., 2019. Anthropogenic carbon as a basal resource in the benthic food webs in the Neva estuary (Baltic Sea). Mar. Pollut. Bull. 146, 190–200.

International Organization for Standardization, 2006. ISO 14040:2006. Environmental Management—Life Cycle Assessment—Principles and Framework. ISO, Geneva.

Johnson, E., 2009. Goodbye to carbon neutral: getting biomass footprints right. Environ. Impact Assess. 29, 165–168.

Kim, J., Park, K., 2018. Effect of the clean development mechanism on the deployment of renewable energy: less developed vs. well-developed financial markets. Energy Econ. 75, 1–13.

Sikdar, S.K., 2003. Sustainable development and sustainability metrics. AICHE J. 49, 1928–1932.

Stamford, L., 2020. Chapter 5—life cycle sustainability assessment in the energy sector. In: Ren, J., Scipioni, A., Manzardo, A., Liang, H. (Eds.), Biofuels for a More Sustainable Future: Life Cycle Sustainability Assessment and Multi-Criteria Decision Making. Elsevier, Amsterdam, pp. 115–163.

United Nations, 2021. Sustainable Development Knowledge Platform. https://sustainabledevelopment.un.org/. accessed December 2021.

United States Environmental Protection Agency, 2012. A Framework for Sustainability Indicators at EPA. https://www.epa.gov/sites/production/files/2014-10/documents/framework-for-sustainability-indicators-at-epa.pdf. accessed December 2021.

Vaz Jr., S., 2020. Treatment of Biomass Residues—A Sustainable Approach. Springer Nature, Cham, pp. 19–27.

World Bank, 2021a. State and Trends of Carbon Pricing 2021. World Bank. World Bank, Washington, DC. https://openknowledge.worldbank.org/handle/10986/35620. (accessed December 2021).

World Bank, 2021b. Understanding CO_2 Emissions from the Global Energy Sector. https://openknowledge.worldbank.org/handle/10986/17143?show=full. accessed December 2021.

World Health Organization, 2018. 9 out of 10 People Worldwide Breathe Polluted Air, But More Countries are Taking Action. https://www.who.int/news/item/02-05-2018-9-out-of-10-people-worldwide-breathe-polluted-air-but-more-countries-are-taking-action. accessed December 2021.

Zheng, J., Suh, S., 2019. Strategies to reduce the global carbon footprint of plastics. Nat. Clim. Chang. 9, 374–378.

CHAPTER 14

Metrics and measurements

Metrics are needed to determine, to quantify, and to prove the economic, environmental, and social impacts that compose sustainability. This is especially true for environmental impacts to demonstrate environmental gains for certain technologies dedicated to renewable carbon (RC) source(s).

Some examples of metrics and their applications in measurement methodologies are discussed in this chapter.

14.1 The lifecycle assessment

To measure the economic, social, and environmental impacts of technologies taking advantage of RC, robust methodologies are needed to quantify and analyze their gains, especially environmental gains.

The lifecycle analysis (LCA) is a sustainability tool that assists in decision-making in various industrial sectors, evaluating the positive and negative impacts of technologies, their processes, and products. LCA studies environmental aspects and potential impacts throughout the life of a product from "cradle to the grave", i.e., from the acquisition of the raw material, through production, use, and disposal of the final product. It is emphasized that resource use, human health, and ecology are the main aspects to be considered during impact assessment.

Historically, the LCA structure took its first steps in the 1960s in the United States with The Coca-Cola Company generating the first standout scheme comparing the environmental performance of its packaging (Kylili et al., 2018).

Currently, LCA is considered the most efficient methodology in the study of environmental impacts and has been highly used in sectors of energy generation, polymer production, mining, transport, etc. The main characteristic of LCA is the possibility of evaluating the environmental impacts of each component of the process, which allows for a more accurate quantification of the impacts of a given route and, consequently, to optimize the proposition of future goals and decision-making.

Renewable Carbon
https://doi.org/10.1016/B978-0-323-99735-5.00009-8

As an example of LCA application for RC, Bishop et al. (2021) explored the growing collective of LCA studies that compare the environmental footprints of specific bioplastics from biomass against those of petrochemical plastics. Forty-four relevant studies published between 2011 and 2020 were reviewed to explore important methodological choices regarding impact category selection, inventory completeness (e.g., inclusion of additives), boundary definition (e.g., inclusion of land-use change impacts), representation of biogenic carbon, choice of end-of-life scenarios, type of LCA, and the application of uncertainty analysis (Fig. 14.1). Good practice examples facilitated identification of common gaps and weaknesses in LCA studies applied to benchmark bioplastics against petrochemical plastics. Many studies did not provide a holistic picture of the environmental impacts of bioplastic products, thereby potentially supporting misleading conclusions. For comprehensive evaluation of bioplastic sustainability, authors recommend that LCA practitioners:

- Embrace more detailed and transparent reporting of lifecycle inventory (LCI) data within plastic LCA studies
- Adopt a comprehensive impact assessment methodology pertaining to all priority environmental challenges
- Incorporate multiple plastic use cycles within functional unit definition and system boundaries where plastics can be recycled
- Include additives in lifecycle inventories unless there is clear evidence that they contribute $<1\%$ to all impact categories
- Apply biogenic carbon storage credits only to long-term carbon sinks
- Account for (indirect) land-use change arising from feedstock cultivation
- Prospectively consider realistic scenarios of deployment and end-of-life, preferably within a consequential LCA framework.

When we consider biomass as the RC source, the land-use change needs to be computed because crops generate this class of impact on the environment with the release of greenhouse gas (GHG)—notably CO_2—as discussed in the Chapter 13 about anthropogenic and biogenic carbons.

Measurement software

Software are indispensable tools to measure the LCA parameters of impacts due to, for example, the data amount (inputs), their complexity, and their results (outputs).

SimaPro[a] is an LCA software that collects, analyzes, and monitors sustainable development data for products and services and is used in

[a] https://simapro.com/.

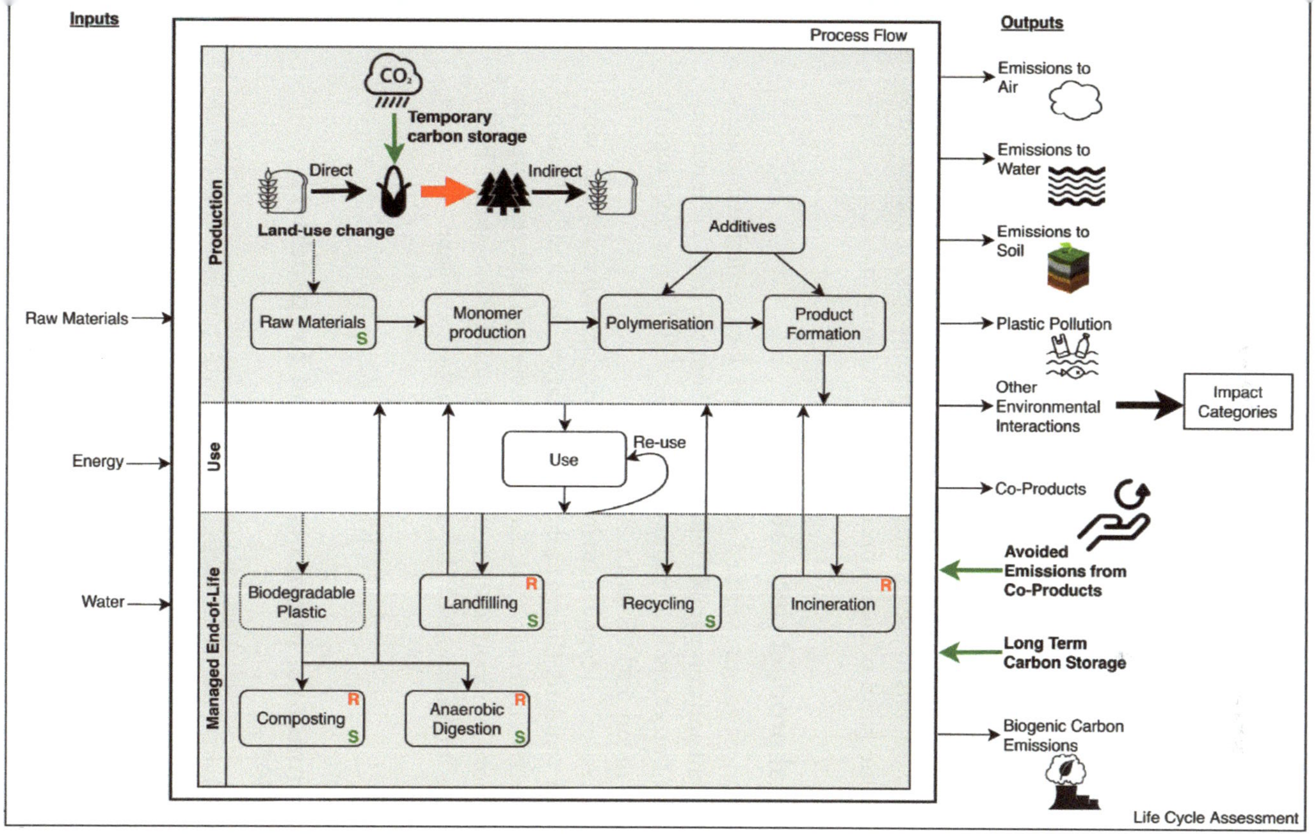

Fig. 14.1 A simplified schematic of a plastic value chain represented in LCA. The main processes, inputs, and outputs are displayed. *Dashed lines* represent flows specific to biodegradable plastics. The diagram also gives an indication of carbon flows throughout the system, with *S* representing carbon storage, and *R* representing the release of the carbon. *(Credit: Bishop, G., Styles, D., Lens, P.N.L., 2021. Environmental performance comparison of bioplastics and petrochemical plastics: a review of life cycle assessment (LCA) methodological decisions. Resour. Conserv. Recycl. 168, 105451. Reprinted with permission from Elsevier.)*

applications such as carbon and water footprints, product design, generation of environmental product declarations (EPD; discussed later), and definition of key performance indicators. SimaPro is the most used LCA software worldwide, covering users and entities in more than 80 countries, including universities, industries, and companies, in addition to following the recommendations of norm ISO 14040:2006 (International Organization for Standardization, 2006a).

Moreover, SimaPro stands out for the possibility of carrying out the analysis by different methods, allowing for the creation of countless scenarios for a given process and comparing its results.

Eco-Indicator 99 is one of the most used methods in SimaPro based on endpoints. It creates ecoindicators that facilitate the modeling of the process, assigning points/weights to each identified impact, allowing for quantitative comparison of the harm of each impact to the environment.

Environmental product declaration

An EPD is an independently verified and registered document that communicates transparent and comparable information about the lifecycle environmental impact of products in a credible way. If the EPD is the final report, the foundation of any EPD is an LCA.

An EPD is a so-called type III environmental declaration compliant with the ISO 14025 standard (International Organization for Standardization, 2006b). A type III environmental declaration is created and registered in the framework of a program, such as the International EPD System.[b] EPDs registered in the International EPD System are publicly available and free to download.

14.2 Environmental impact calculation for products and processes

Through an environmental impact calculation, it is possible to prove an environmental gain for a certain technology for conversion of RC raw materials. LCA assesses environmental impacts and potential impacts through:

- Compilation of an inventory of inputs and outputs from a product system, the LCI

[b] https://www.environdec.com/home.

- Assessing the potential environmental impacts associated with these inputs and outputs
- Interpreting the results of the inventory analysis and impact assessment phases in relation to the objectives of the studies

As introduced in Chapter 13, norm ISO 14040:2006 (International Organization for Standardization, 2006a) systemizes the framework of LCA analysis by means of the following steps:

(i) Goal definition
(ii) Scope definition
(iii) Inventory analysis
(iv) Impact assessment
(v) Interpretation

The first and second steps define the goal and scope of the LCA. These steps are intended to ensure that the LCA will provide the expected results. The intended application, the reasons for conducting the study, and the target audience must be consistently defined.

The third step, which corresponds to inventory analysis, involves data collection and calculation procedures to quantify the inputs and outputs of a production system.

In the fourth step, impact assessment, the results are integrated; it is at this stage that the potential impacts of a production system are understood and evaluated. This integration takes place from the data collected in the inventory.

Finally, in the fifth step, interpretation corresponds to the phase of analyzing the results of the inventory and impact assessment to assess whether the combinations are consistent with the goals and scope of the LCA. This phase allows conclusions to be drawn to form future recommendations.

As a tool in constant development, LCA still has many uncertainties and limitations. A high demand on time, financial, and human resources are the main limitations as they make it difficult to collect inventory data, as they require high precision.

The absence of data can influence the final result and, consequently, the reliability of the entire LCA. Therefore, it is essential to analyze cost-benefit, time, and other demands added to an LCA, especially because it is an analysis of potential impacts and not actual environmental impacts, i.e., the analysis does not quantify the real value of the impact on the environment.

The limitations must be considered to better estimate the projections resulting from the interpretation of the results, but they do not invalidate the LCA.

Fig. 14.2 depicts a logical sequence considering the LCI and the lifecycle impact assessment (LCIA), components of the LCA study, for several situations. In a summarized way, situations are defined as (European Commission—Joint Research Centre—Institute for Environment and Sustainability, 2010):

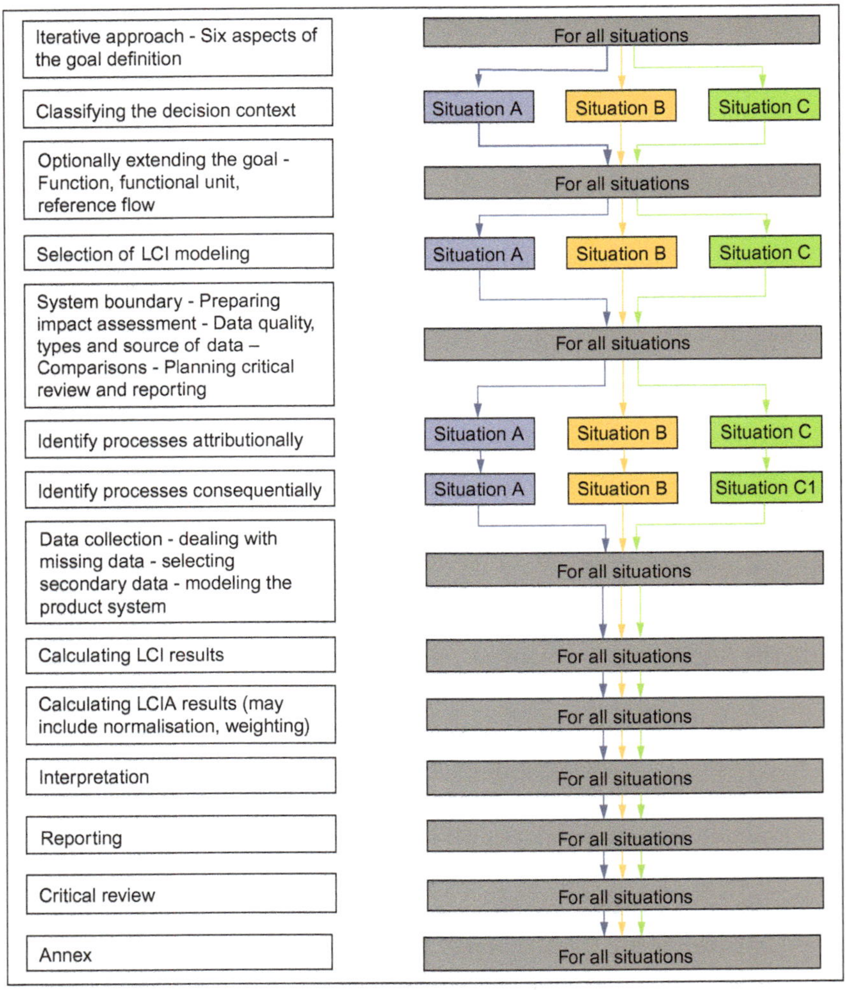

Fig. 14.2 Different goals situation for a LCA study. *LCI*, lifecycle inventory; *LCIA*, lifecycle impact assessment. *(Source: European Commission—Joint Research Centre—Institute for Environment and Sustainability, 2010. International Reference Life Cycle Data System (ILCD) Handbook—General guide for Life Cycle Assessment—Provisions and Action Steps. EUR 24378 EN. Publications Office of the European Union, Luxembourg. Reprinted with permission from European Commission.)*

- Situation A: this comprises microlevel, product, or process-related decision support studies. The lifecycle is modeled by depicting the existing supply-chain, i.e., attributionally.
- Situation B: this comprises mesolevel and macrolevel, strategic ("policy") decision support studies. The analyzed systems shall be modeled as in Situation A, except for those processes in the background system that are affected by large-scale consequences of the analyzed decision.
- Situation C: most monitoring type studies fall into Situation C1; Situation C2 studies are less common. For Situation C1, the lifecycle and all cases of multifunctionality are modeled as in Situation A.

Carbon quantification for certification

Carbon quantification is the basis for the carbon footprint and LCA report certification by a third party, i.e., a certifying body.

GHGs, i.e., CO_2, CH_4, N_2O, and fluorinated gases, present in the atmosphere (outdoor) or in a closed environment (indoor) can be measured by means of sensors and probes based on, for example, infrared radiation.[c] The GHG quantification—especially for CO_2 and CH_4—supply input data to all sustainability measurement approaches.

A GHG concentration can be measured as:

✓ Parts per million (ppm), e.g., $mg\,m^{-3}$
✓ Parts per billion (ppb), e.g., $\mu g\,m^{-3}$
✓ And even parts per trillion (ppt), e.g., $ng\,m^{-3}$

The standardized analytical methodology for carbon quantification can be used according to—or adapted from—those norms preconized by the International Organization for Standardization (ISO), the norm ISO/TS 14067 (International Organization for Standardization, 2013), and orientations from Technical Committee 265 (International Organization for Standardization, 2021).

[c] https://www.gasmet.com/applications/emissions-monitoring/?utm_term=emission%20monitoring&utm_source=adwords&utm_campaign=Emissions+Monitoring+Pillar+Page&utm_medium=ppc&hsa_kw=emission%20monitoring&hsa_ad=516192609099&hsa_grp=124197864667&hsa_ver=3&hsa_acc=2182609276&hsa_cam=12805397706&hsa_mt=b&hsa_src=g&hsa_net=adwords&hsa_tgt=kwd-296288165976&gclid=Cj0KCQiAkNiMBhCxARIsAIDDKNUAT3g2elQ_5zcTZGU_b8xTbWjc5OIJ_Dd4p0fw-ORfVrIRqQtYTBMaAoUzEALw_wcB.

14.3 LCA for biofuels: A case study

This case study is based on the Brazilian program RenovaBio (Brasil, 2020) for biofuels, e.g., ethanol and biodiesel, with objectives to:

- ✓ Provide an important contribution to the fulfillment of the commitments determined by Brazil under the Paris Agreement
- ✓ Promote the adequate expansion of biofuels in the energy matrix, with an emphasis on regular fuel supply
- ✓ Ensure predictability for the fuel market, inducing gains in energy efficiency and reduction of GHGs emissions in the production, sale, and use of biofuels

Fuel distributors must prove compliance with mandatory individual targets through the purchase of decarbonization credits (CBIOs), a financial asset negotiable on the stock exchange, derived from the certification of the biofuel production process based on the respective levels of efficiency achieved in relation to their emissions.

RenovaCalc's methodology (developed by the Brazilian Agricultural Research Corporation (2017)), quantifies GHG emissions over the lifecycle of biofuels, generating a performance index in $gCO_2eq\ MJ^{-1}$ of biofuel. Therefore, those that present production processes with less carbon emissions will have access to CBIOs.

RenovaCalc is a tool that works as a calculator to prove the environmental performance of biofuel production by biofuel industrial plants, where they must detail agricultural and industrial aspects of their production processes that result in carbon emissions, relating energy efficiency and GHG emissions—based on LCA—and establishing guidelines for its certification. The total emission is compared with that of an equivalent fossil fuel (gasoline, in the case of ethanol, or diesel, for biodiesel) resulting in a final score, characterizing the mitigation of emissions. This assigned note becomes a multiplier factor at the time of issuance of CBIOs traded on the stock exchange and that will act as a new product for companies.

In the RenovaBio Program, the performance of biofuels in terms of GHG emissions is quantified in $gCO_2eq\ MJ^{-1}$ by the LCA, although a full LCA study should cover several categories of environmental impact related to the protection of natural resources, ecological systems, and human health. In the RenovaBio Program, it is considered only in the "climate change" category.

RenovaBio adopts attributional LCA that aims to attribute to a product, manufactured at a given time, a portion of the total pollutant emissions and consumption of resources in the economy (Sonnemann and Vigon, 2011).

The range of calculations is "from the cradle to the grave". In this scope, all material and energy flows are considered: those consumed by production processes and emitted to the environment, the extraction of natural resources, acquisition or production and treatment of biomass, its conversion into biofuel, and its combustion in engines, including all transport phases.

Biofuels planned in the first phase of RenovaBio, whose production routes are already included in the RenovaCalc, are:

- First-generation sugarcane ethanol
- First- and second-generation ethanol integrated in plant
- Second-generation ethanol in a dedicated plant
- Sugarcane and corn ethanol in an integrated mill ("flex")
- Corn ethanol in a dedicated plant
- Imported corn ethanol
- Biodiesel
- Aviation biokerosene by HEFA (hydroprocessed esters and fatty acids) from soybeans
- Biomethane from waste

RenovaCalc structure

For each biofuel route, general identification data of the Production Unit are requested, as well as information on compliance with program eligibility criteria (related to control measures to avoid the suppression of native vegetation), and data on the production process, which are distributed into:

(a) Agricultural phase (when relevant)
(b) Industrial phase
(c) Distribution phase

RenovaCalc accounts for emissions based on the information of the agricultural and industrial phase provided by biofuel producers, generating the intensity index of biofuel carbon, which is subsequently subtracted from the corresponding fossil fuel index, generating the energy-environmental efficiency note (in $gCO_2eq\ MJ^{-1}$).

The agricultural phase contributes significantly to GHG emissions from biofuels, which are related mainly to the production and use of inputs. The most important GHGs generated in agricultural activity are methane (CH_4), carbon dioxide (CO_2), and nitrous oxide (N_2O). At this stage, the main practices that impact emissions of GHGs are the use of limestone, the use of nitrogenous inputs, the burning of agricultural residue, and the consumption of fossil fuel in mechanized operations, in addition to land use change. Contributions arising from production and agricultural use of pesticides are of little significance on its impact on climate change, which is considered in RenovaBio. Therefore, these inputs are considered in the accounting of carbon intensity of biofuels in the form of standard data (not penalized, corresponding to the common practice today for each agricultural crop).

There are several models for estimating GHGs emissions from agricultural production processes, with different levels of complexity and specificity. The most widespread models propose the accounting of substances generators and their multiplication by specific emission factors for each type of GHG. The main methodological references for estimating these emissions in RenovaBio are the IPCC Guides, in particular v.4 "Agriculture, Forestry and Other Land Use" (Intergovernmental Panel on Climate Change, 2006a). When available, specific emission factors for the region and culture under analysis were preferred to the standard issue ("default") of the IPCC.

The contribution of industrial stage processes to the environmental performance of the biofuel lifecycle is mostly associated with the yield of product(s) and co-product(s) and the consumption of fuel and energy (electricity). Emissions arising from the production of chemical inputs used in industrial processes and waste treatment are usually not significant for the main biofuels currently produced in a large scale in the country, as these are considered in the accounting of the carbon intensity of biofuels based on the typical profile of each technological route. However, for corn ethanol, second-rate ethanol generation, and biodiesel, some industrial inputs can significantly contribute to its intensity and these are used as input parameters in RenovaCalc.

For industrial processes, GHGs emission estimates are based on the IPCC Guide, in particular v.3 "Industrial Processes and Product Use" (Intergovernmental Panel on Climate Change, 2006b). Again, when available, specific emission factors for the processes under analysis were preferred to the default emission factors ("default") of the IPCC.

After filling in all RenovaCalc parameters, the tool calculates the carbon intensity of the biofuel, generating an index in "$gCO_2eq\ MJ^{-1}$." This process is done automatically on the calculator, using six main steps:

(i) Adaptation of the input parameters to a reference flow (ton of biomass processed) and insertion as "input flows" in the LCI of each production route;

(ii) Linking these inventory "inflows" to GHGs emissions data upstream of the agroindustrial process related to the production of agricultural and industrial inputs and energy generation, and downstream related to the distribution and use of biofuel

(iii) Consolidation of an inventory of generated GHGs emissions in the agricultural, industrial, distribution, and use of biofuel phases

(iv) Conversion of GHGs emissions to unit "gCO_2eq" based on characterization factors for each gas, according to GWP100—the global warming potential for 100-year horizon (United States Environmental Protection Agency, 2021)

(v) Adequacy of the index to the functional unit (MJ of biofuel);

(vi) Calculation of carbon intensity difference for the biofuel with its fossil fuel reference, generating the final grade

As an example of calculated value by means the RenovaCalc[d], we have 26.8 $gCO_2eq\ MJ^{-1}$ for biodiesel.

14.4 Conclusions

Lifecycle assessment is highlighted as the main tool for sustainability measurements for several processes and products related to RC. Indeed, LCA is considered the most efficient methodology in the study of environmental impacts and has been highly used in sectors of energy generation, polymer production, mining, transport, biomass, etc. However, limitations must be considered to better estimate the projections resulting from the interpretation of the results; however, they do not invalidate the LCA.

Thus, the application of calculation tools based on LCA promotes carbon mitigation by means of carbon quantification and pricing, as demonstrated for biofuels.

References

Bishop, G., Styles, D., Lens, P.N.L., 2021. Environmental performance comparison of bioplastics and petrochemical plastics: a review of life cycle assessment (LCA) methodological decisions. Resour. Conserv. Recycl. 168, 105451.

Brasil, 2020. RenovaBio. https://www.gov.br/anp/pt-br/assuntos/renovabio. accessed December 2021.

Brazilian Agricultural Research Corporation, 2017. RenovaCalc. https://www.embrapa.br/busca-de-solucoes-tecnologicas/-/produto-servico/4355/renovacalc- - -ferramenta-para-comprovacao-do-desempenho-ambiental-de-usinas-de-biocombustiveis. accessed December 2021.

European Commission—Joint Research Centre—Institute for Environment and Sustainability, 2010. International Reference Life Cycle Data System (ILCD) Handbook—

[d] https://www.gov.br/anp/pt-br/assuntos/renovabio/renovacalc.

General guide for Life Cycle Assessment—Provisions and Action Steps. EUR 24378 EN. Publications Office of the European Union, Luxembourg.
Intergovernmental Panel on Climate Change, 2006a. IPCC Guidelines for National Greenhouse Gas Inventories: V. 4—Agriculture, Forestry and Other Land Use. Prepared by the National Greenhouse Gas Inventories Programme, Eggleston HS, Buendia L, Miwa K, Ngara T, Tanabe K. Japan: IGES.
Intergovernmental Panel on Climate Change, 2006b. IPCC Guidelines for National Greenhouse Gas Inventories: V. 3—Industrial Processes and Product Use. Prepared by the National Greenhouse Gas Inventories Programme, Eggleston HS, Buendia L, Miwa K, Ngara T, Tanabe K. Japan: IGES.
International Organization for Standardization, 2006a. ISO 14040:2006. Environmental Management—Life Cycle Assessment—Principles and Frameworks. ISO, Geneva.
International Organization for Standardization, 2006b. ISO 14025:2006. Environmental Labels and Declarations—Type III Environmental Declarations—Principles and Procedures. ISO, Geneva.
International Organization for Standardization, 2013. ISO/TS 14067. Greenhouse Gases—Carbon Footprint of Products—Requirements and Guidelines for Quantification and Communication. (Technical Specifications). International Organization for Standardization, Geneva.
International Organization for Standardization, 2021. ISO/TC 265. Carbon Dioxide Capture, Transportation and Geological Storage. https://iso.org/committee/648607.html. (accessed December 2021).
Kylili, A., Seduikyte, L., Fokaides, P.A., 2018. Chapter 9—life cycle analysis of polyurethane foam wastes. In: Thomas, S., Rane, A.V., Kanny, K., Abitha, V.K., Thomas, M.G. (Eds.), Plastics Design Library, Recycling of Polyurethane Foams. William Andrew Publishing, pp. 97–113.
Sonnemann, G., Vigon, B., 2011. Global Guidance Principles for Life Cycle Assessment Database—"Shonan Guidance Principles". SCP documents. Vol. 158 UNEP–SETAC, Geneva.
United States Environmental Protection Agency, 2021. Understanding Global Warming Potentials. https://www.epa.gov/ghgemissions/understanding-global-warming-potentials. accessed December 2021.

CHAPTER 15

Alternative sources to petrochemicals: New value chains

The petrochemical industry is a technological and economic stakeholder that grew in the 20th century with a deep impact on the modern society because oil-derived products are present every day, from polymers and fuels to pharmaceuticals, among various others.

However, petrochemicals are responsible for several pollution and contamination issues that create the need for new sustainable value chain alternatives to these end products as discussed in this chapter.

15.1 Introduction to the petrochemical value chains

There are no doubts that petrochemical products are everywhere and are integrated in modern society. They include plastics, fertilizers, packaging, clothing, digital devices, medical equipment, detergents, tires, and many others. They are also found in many parts of the modern energy system, including solar panels, wind turbine blades, batteries, thermal insulation for buildings, and electric vehicle parts.

According to the International Energy Agency (2021), petrochemicals are already a major component of the global energy system, and its importance is continuing to grow. The demand for plastics—the most familiar group of petrochemical products—has outpaced that of all other bulk materials (such as steel, aluminum, or cement) and has nearly doubled since 2000. Advanced economies, such as the United States and Europe, currently use up to 20 times as much plastic and up to 10 times as much fertilizer as developing economies, such as India and Indonesia, on a per capita basis. This underscores the huge potential for worldwide growth. Other petrochemicals include:

- Plastic packaging for food and other commercial products can be made from a range of petrochemical products, including polyethylene and polystyrene

https://doi.org/10.1016/B978-0-323-99735-5.00016-5

- Globally, more than a half of ammonia is converted into urea, which is, in turn, used as a fertilizer to increase crop yields and boost food production
- Synthetic rubber is a major component of tires for cars, trucks, and bicycles, and is mainly derived from the petrochemical butadiene
- Many laundry detergents and items of clothing in our washing machines are derived from petrochemicals, such as surfactants and polyester fiber

The growth in demand for petrochemical products means that petrochemicals are set to account for more than one-third of the growth in oil demand to 2030 and nearly half to 2050, ahead of trucks, aviation, and shipping. Petrochemicals are also poised to consume an additional 56 billion cubic meters of natural gas by 2030, equivalent to about half of Canada's total gas consumption today (International Energy Agency, 2021).

Figs. 15.1 and 15.2 describe this demand, while Figs. 15.3 and 15.4 depict the detailed petrochemical chains.

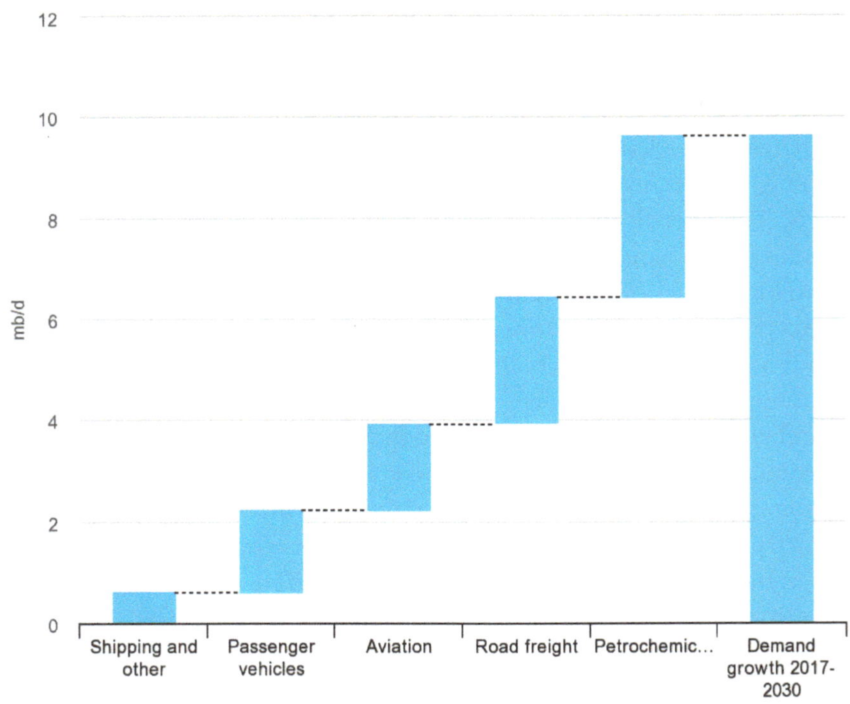

Fig. 15.1 Oil demand growth by sector, 2017–30. *(Credit: International Energy Agency (https://www.iea.org/data-and-statistics/charts/oil-demand-growth-by-sector-2017-2030). Reprinted with permission from International Energy Agency.)*

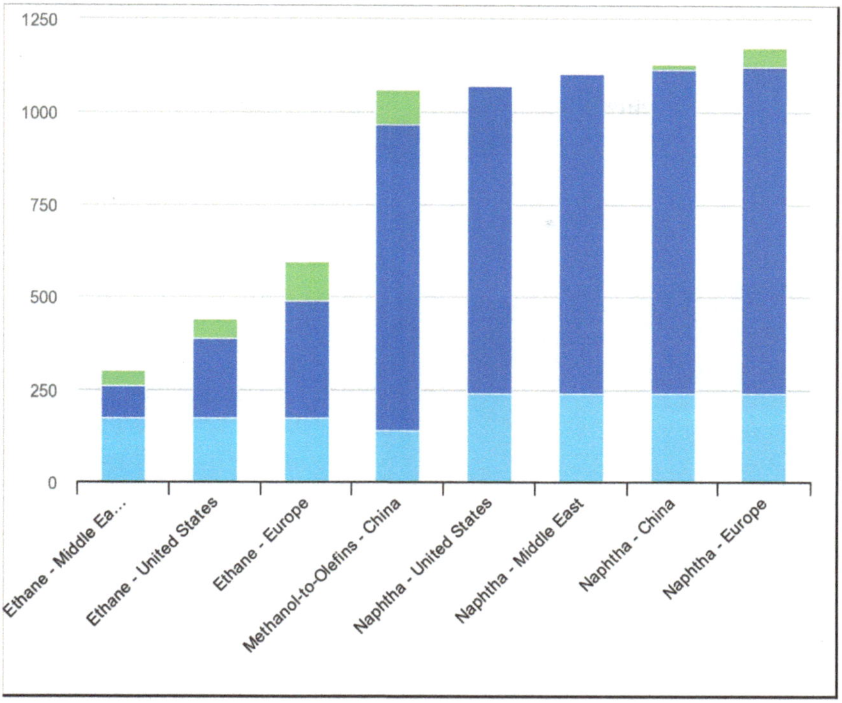

Fig. 15.2 Simplified levelized cost of petrochemicals for selected feedstocks and regions, 2017; *x* axis represents USD/per tonne of high value chemicals (HVC). *Green*, process energy; *dark blue*, feedstock; *light blue*, CAPEX/OPEX. *(Credit: International Energy Agency (https://www.iea.org/data-and-statistics/charts/simplified-levelised-cost-of-petrochemicals-for-selected-feedstocks-and-regions-2017). Reprinted with permission from International Energy Agency.)*

Despite their relevance for the global economy, petrochemicals are a potential source of pollution of the ecosystems and contamination of people and animals due to oil exploration, processing, and end use of their related products.

As an example of this pollution, there are oil spills that are, generally, the result of accidents at oil wells or on pipelines, ships, trains, and trucks that move oil from wells to refineries (United States Energy Information Administration, 2021). Oil spills pollute soil and water and may cause devastating explosions and fires. Regarding human contamination, it is worth citing the presence of plastic particles (i.e., microplastics and nanoplastics) in leaving organisms with induced cellular toxicity in mammals (Banerjee and Shelver, 2021); moreover, the polyaromatic hydrocarbons (PAHs, Fig. 15.5) are recognized lung cancer agents (Moorthy et al., 2015).

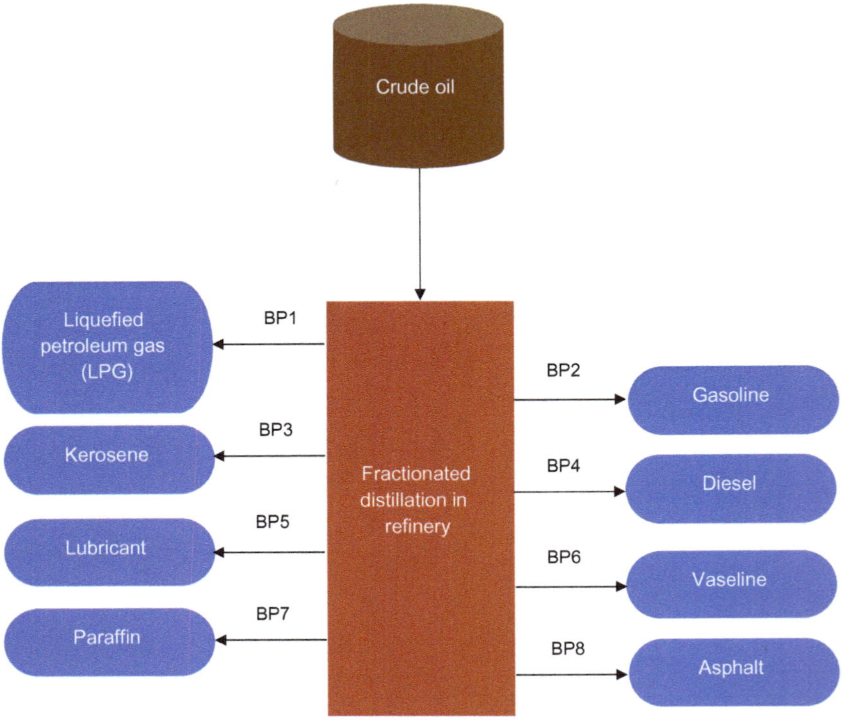

Fig. 15.3 Petrochemical value chains from the oil distillation according to the increasing in the boiling point (from LPG to asphalt); BP: boiling point (in °C); lower BP, higher volatility and higher vapor pressure. *(Credit: Author.)*

From these observations, it is necessary to search for and develop sustainable value chains to substitute petrochemicals—all or in part.

15.2 Alternative feedstocks for chemicals and materials

Renewable carbon (RC) raw materials are growing as alternative and sustainable feedstocks for chemicals and materials derived from oil, going straight to the principle 7 of green chemistry (use of renewable feedstocks), especially in the case of biomass components; moreover, CO_2 can also be considered a renewable feedstock from this perspective.

As discussed in Part 2, several technologies based on chemical, biochemical, and thermochemical routes can be applied to convert RC raw materials into chemicals, materials, and biofuels, among other possibilities of products that promote the development of new value chains.

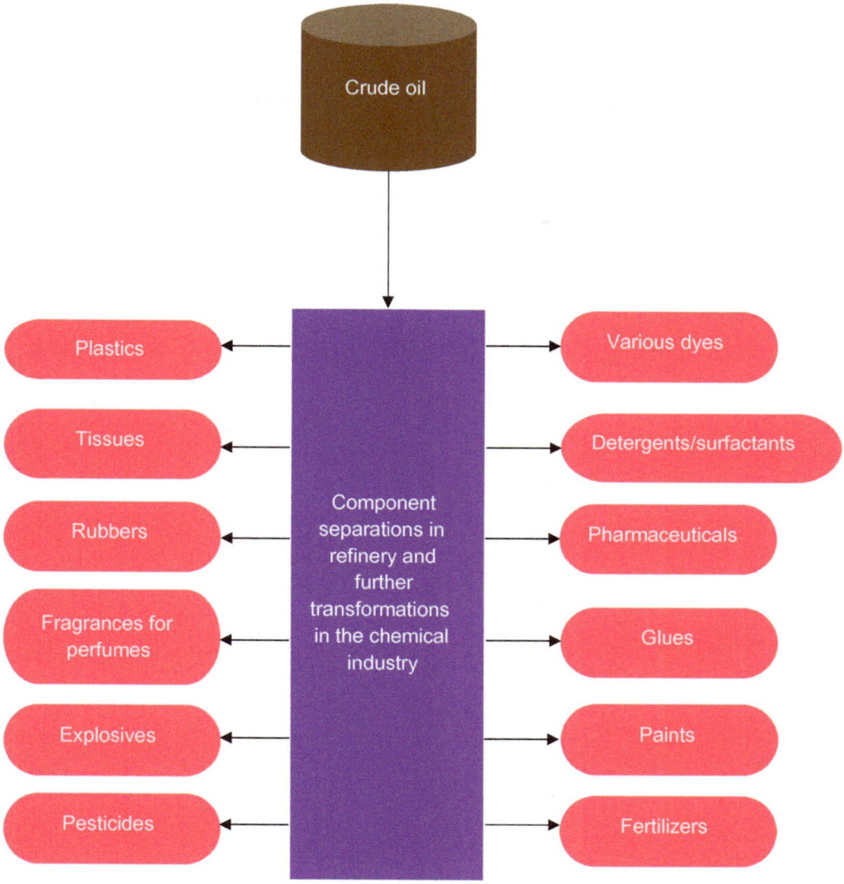

Fig. 15.4 Petrochemical value chain from oil components transformation. *(Credit: Author.)*

These alternative feedstocks can be classified according to the end products (or partial end products) as described in Table 15.1.

Aromatics, e.g., benzene, toluene, ethylbenzene, and xylenes (BTEX), can be obtained by means of the catalytic cracking of lignin (Ryu et al., 2019), which can be a substitute to oil-derived BTEX, although they are toxic and ecotoxic compounds (Cheng et al., 2015).

Of course, it is not so simple to substitute petrochemicals in modern society due their economic and industrial relevance. However, the use of renewable or green products can confer a desirable sustainability approach to those value chains based on RC raw materials with valorization of these productive systems.

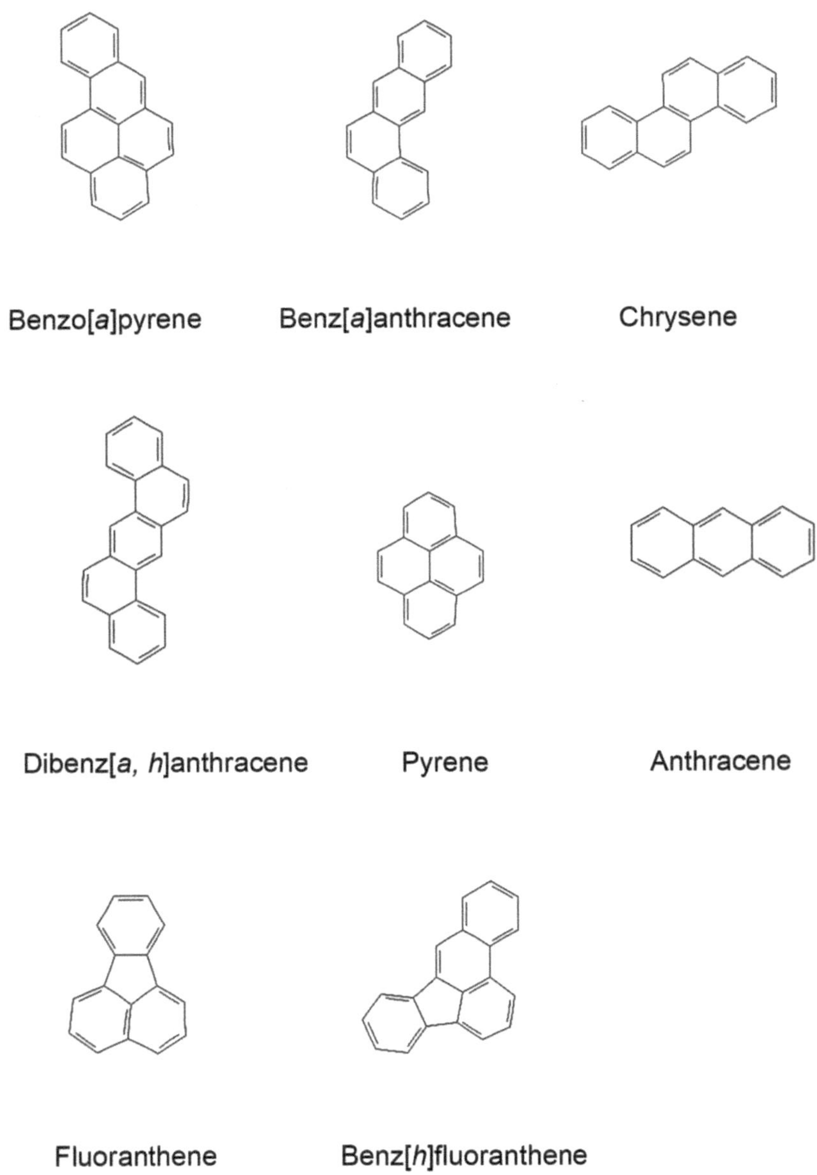

Fig. 15.5 Polyaromatic hydrocarbons (PAHs) derived from incomplete oil combustion known as carcinogenic agents. *(Credit: Author.)*

Table 15.1 Alternative feedstocks for chemicals, fuels, and materials as petrochemical substitutes.

Product	Application	Feedstock	Renewable carbon category
Biodiesel (esters)	Biofuel	Vegetable oil	Biomass
Ethanol	Biofuel	Sucrose	Biomass
Biogas (methane)	Biofuel	Organic wastes	Biomass
Carbonates	Agriculture, several industry sectors	CO_2	CO_2
Cyclic carbonates	Polymer precursors, fuel additives, or electrolytes in batteries, and aprotic high-boiling point solvents	CO_2	CO_2
Alcohols	Polymer precursors, solvents, biofuels, sanitizer, among others	Sugars	Biomass
Organic acids	Polymer precursors, food and feed additives, among others	Sugars	Biomass
Aldehydes	Precursors for several syntheses, flagrance, agrochemical, solvents, among others	Sugars	Biomass
Lignosulphonates	Flocculant, agrochemical formulations, dispersant	Lignin	Biomass
Aromatics	BTEX for fuels	Lignin	Biomass
Phenols	Resins	Lignin	Biomass
Monomers	Plastics	Waste	Recyclable materials

15.3 Strategy to create sustainable value chains based on renewable carbon

A strategy to create or develop sustainable value chains based on RC sources should consider some aspects:

- Is there a well-established production chain?
- Is there a certification of sustainability for the product(s)?
- Will the product(s) be used as substitute (drop-in) to a petrochemical?

A well-established production will help develop a predictable market for a green product, while a certification of sustainability will facilitate market acceptance.[a] A product to be used as a substitute for a certain petrochemical (drop-in) can replace conventional chemicals without any need for changes in the existing infrastructure because it is chemically identical to the oil-derived product. Then this strategy can be more easily put into practice than others that involve the need to change the infrastructure of production. On the other hand, in contrast to drop-in chemicals, "dedicated biobased chemicals" (Nova Institute, 2017) have no fossil counterpart but represent renewable products with unique properties presenting a necessity to develop infrastructure and market that will increase costs and risks.

As an innovative strategy, the insertion of concepts presented in Parts 1 and 2—bioeconomy, circular economy, and green chemistry—will certainly help to overcome challenges and aggregate value.

Moreover, the application of lifecycle analysis (LCA) and its related parameters, according to discussions in Chapter 14, e.g., carbon footprint, will boost new value chains. For example, this type of assessment for value chains was described by Meiinrenken et al. (2020) concluding that a sector-level understanding of emissions, absent of individual product carbon footprints (PCFs), is insufficient to reliably quantify carbon emissions.

15.4 Renewable building blocks: A case study

This case study is based on the work of Pellis et al. (2016) regarding the production of renewable building blocks from biomass for sustainable polyesters by means of biotechnological routes (Fig. 15.6).

The next generation of plastics are expected to contribute to a massive reduction in the carbon footprint by the exploitation, in industrial productive processes, of renewable monomers such as polyols and dicarboxylic acids obtainable via biotechnological production. More specifically, there

[a] Greenhouse gases (GHGs) emissions verification and carbon footprint: https://www.sgs.com/-/media/global/documents/infographics/esg-certification-infographic-en.pdf.

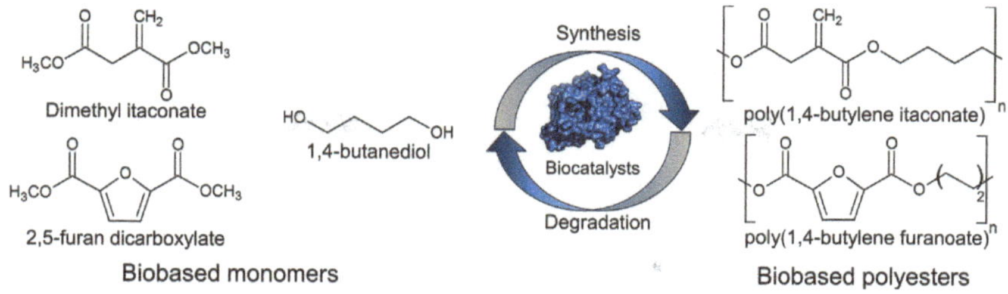

Fig. 15.6 Bioanalyzed synthesis and recycling of aliphatic and aromatic renewable polyesters: the use of enzymes as catalysts confers biodegradability to the produced polymers. *(Reprinted from Pellis, A., Acero, E.H., Gardossi, L., Ferrario, V., Guebitz, G.M., 2016. Renewable building blocks for sustainable polyesters: new biotechnological routes for greener plastics. Polym. Int. 65, 861–871, with permission from Wiley.)*

is a rising demand for advanced polyesters displaying new functional properties while meeting higher sustainability criteria.

Polyesters are part of everyday life with applications in clothing, food packaging, car manufacturing, and biomedical devices.

The following biobased monomers can be obtained by means of biotechnological routes to reach biobased polyesters:

- Itaconic acid
- Adipic acid
- Succinic acid
- Terephthalic acid
- Furandicarboxylic acid
- 1,3-Propanediol
- 2,3-Butanediol
- 1,4-Butanediol
- Glycerol
- Sorbitol and isosorbide
- Ethylene glycol

Thus, this example demonstrates the feasibility of petrochemical substitution by biobased chemicals and materials.

15.5 Conclusions

There are no doubts that petrochemical products are everywhere and are integrated in modern society. They include plastics, fertilizers, packaging, clothing, digital devices, medical equipment, detergents, tires, and many others. They are also found in many parts of the modern energy system, including solar panels, wind turbine blades, batteries, thermal insulation for buildings, and electric vehicle parts. Indeed, oil demand as feedstock is growing on a global scale.

However, the petrochemical industry is a source of pollution and contamination, requiring alternative feedstocks and technological strategies to reach its sustainability as renewable building blocks.

References

Banerjee, A., Shelver, W.L., 2021. Micro- and nanoplastic induced cellular toxicity in mammals: a review. Sci. Total Environ. 755 (Part 2), 142518.

Cheng, P., Lee, J.-W., Sichani, H.T., Ng, J.C., 2015. Toxic effects of individual and combined effects of BTEX on *Englena gracilis*. J. Hazard. Mater. 284, 10–18.

International Energy Agency, 2021. The Future ofPetrochemicals—Towards a More Sustainable Chemical Industry. https://www.iea.org/reports/the-future-of-petrochemicals. (Accessed December 2021).

Meiinrenken, C.J., Chen, D., Esparza, R.A., Iyer, V., Paridis, S.P., Prasad, A., Whillas, E., 2020. Carbon emissions embodied in product value chains and the role of life cycle assessment in curbing them. Sci. Rep. 10, 6184.

Moorthy, B., Chu, C., Carlin, D.J., 2015. Polycyclic aromatic hydrocarbons: from metabolism to lung cancer. Toxicol. Sci. 145, 5–15.

Nova Institute, 2017. Which Implementation Strategy is Right for Bio-Based Chemicals—Drop-in, Smart Drop-in and Dedicated Ones? https://renewable-carbon.eu/news/which-implementation-strategy-is-right-for-bio-based-chemicals/. (Accessed December 2021).

Pellis, A., Acero, E.H., Gardossi, L., Ferrario, V., Guebitz, G.M., 2016. Renewable building blocks for sustainable polyesters: new biotechnological routes for greener plastics. Polym. Int. 65, 861–871.

Ryu, H.W., Lee, H.W., Jae, J., Park, Y.-K., 2019. Catalytic pyrolysis of lignin for the production of aromatic hydrocarbons: effect of magnesium oxide catalyst. Energy 179, 669–675.

United States Energy Information Administration, 2021. Oil and Petroleum Products Explained—Oil and the Environment. https://www.eia.gov/energyexplained/oil-and-petroleum-products/oil-and-the-environment.php. (Accessed December 2021).

CHAPTER 16

Strategies for industrial ecology

Industrial ecology (IE) is a new and comprehensive field of study that advocates an integrated approach to the relationship between industry and the environment.

In a general way, the proposal is that the resources used by the industry remain within the production cycle, thus avoiding waste. Then IE meets the tenets of circular economy and green chemistry, as discussed in this chapter.

16.1 Introduction to industrial ecology

The concept of IE emerged in the wake of the environmental movements that intensified in the 1970s and that took shape with the United Nations Environment Program.[a]

We can consider IE as a scientific branch closely related to sustainability because it proposes the development of production systems considering the mimetization of the dynamic nature for ecosystems that can promote positive environmental impacts on those processes and products based on renewable carbon (RC) sources, as discussed in Part 2. Moreover, IE is related to the concept of sustainability insofar as it involves closed production cycles, with reduction or elimination of dependence, for example, on nonrenewable energy sources with such aspects generating positive impacts on the environment, on the local economy and, consequently, on society.

From these definitions, IE is adequately related to the circular economy by means of a strong contribution to the recycling approach, as depicted in Fig. 16.1.

From Fig. 16.1, we can observe two categories of pathways:

✓ Disposal pathways—not related to a circular economy because raw materials and products are not reused, and such paths are undesirable.

✓ Recycling and reuse pathways—directly related to a circular economy because raw materials and products are reused, and such paths are stimulated by RC, i.e., biomass waste and recycling materials.

[a] https://www.undp.org/.

Renewable Carbon
https://doi.org/10.1016/B978-0-323-99735-5.00002-5

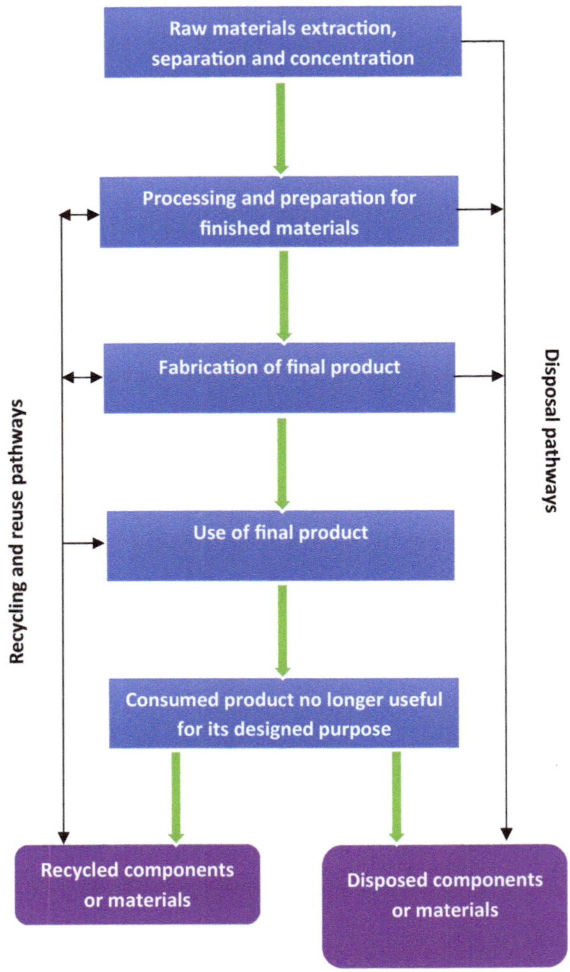

Fig. 16.1 Flowchart of a complete industrial ecosystem. *(Credit: author.)*

The assessment of social, economic, and environmental aspects for products and processes, as preconized by the sustainability concept discussed in Chapter 13, is a contribution to establish the IE concept, as depicted by Fig. 16.2.

16.2 Ecofriendly approaches for renewable carbon

First, we can highlight that industrial symbiosis (IS)—part of the holistic view of IE—is the process by which wastes or by-products of an industry

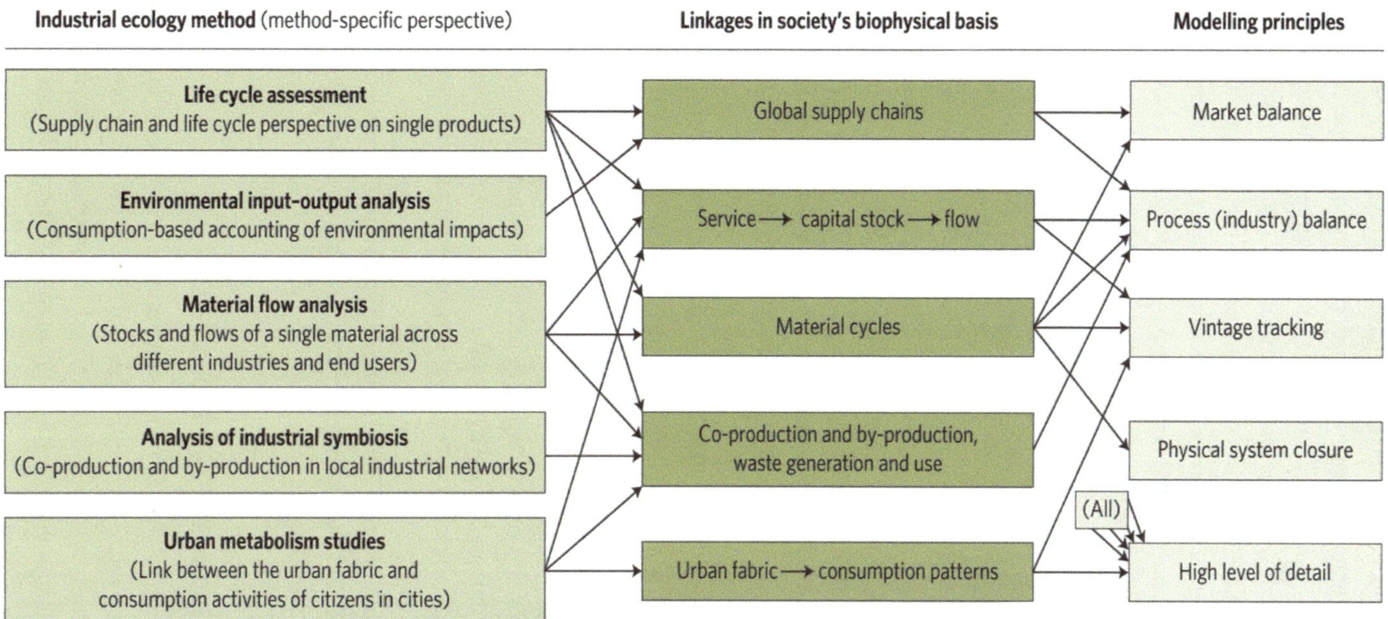

Fig. 16.2 Overview of the main descriptive and assessment methods in industrial ecology. *(Credit: Paulik, S., Arvesen, A., Stadler, K., Hertwich, E.G., 2017. Industrial ecology in integrated assessment models. Nat. Clim. Chang. 7, 3–20. Reprinted with permission from Springer Nature.)*

or industrial process become raw materials for another (European Commission, 2018). In a practical way, IS can support to create the circular economy. Then we can apply principle 7 of green chemistry (use of renewable feedstocks) for the processing of RC raw materials.

One of the pioneering cases of application of the concept and principles of IE is the cluster of industries in the municipality of Kalundborg, Denmark.[b] There, a group of nine large companies operates in a system in which more than 30 products are exchanged between them. For example, excess gas from an oil refinery process is treated to remove sulfur, which in turn is used to make sulfuric acid, which is used by two other industries in the cluster.

In another case, the residue from the production of a pharmaceutical industry, in the form of biomass (a RC raw material), is used as fertilizer by farms in the region. Kalungborg's urban symbiosis has influenced parks around the world, from the United States to Japan.

Additionally, and according to Francis (2003), the cleaner production approach motivated by the IE looks for ways to reduce the risks to humans and the environment arising from production processes by conserving raw materials and energy, eliminating toxic raw materials, and reducing the quantity and the toxicity of all emissions and wastes. Furthermore, it is possible to observe that this approach maintains high similarities with green chemistry principles and converges industrial and economic paradigms, which contribute to sustainability.

Giurco et al. (2014) stated an interesting observation that if a carbon property right for soil carbon emerges before a property right for subterranean carbon, then biobased IE opportunities—a model based on RC—could be enabled ahead of a regional symbiosis involving carbon capture and storage.

Finally, Baldassare et al. (2019) studied IS from two perspectives: circular economy and IE, and found that they are necessary to define an IS cluster, as in the case of Kalundborg.

Software for industrial symbiosis

Nowadays, there is software to assess the potential IS—one of those methods is presented in Fig. 16.2—in light of IE with IS comprising the co-production and by-production related to waste generation and its use for the industrial process balance (e.g., mass and energy balance).

[b] http://www.symbiosis.dk/en/.

This type of software can support decision making in two areas (FISSAC Project, 2020):

(i) Lifecycle approach to material flow analyses and industrial clustering providing:
- ➢ Accurate and dynamic lifecycle assessment (LCA) of the environmental impact and material flow costs of proposed IS solutions
- ➢ Quantification of economic benefits in a holistic manner
- ➢ Dynamic/live by-product and waste flow analyses on a map (georeferencing of facilities)

(ii) GIS[c]-based approach for:
- ➢ Geo-referencing waste flows: publishing data on waste availability and characteristics on a map
- ➢ Boosting the impact of results produced by each IS system on the platform: facilitating contacts among industrial players of the construction value chain and encouraging research and development for underdeveloped waste streams
- ➢ Quantifying economic and environmental burdens in relation to LCA capabilities

Thus, IE can be applied to the industrial process in a systematic way.

16.3 Industrial ecology in chemical industry: A case study

This case study is based on the work of Yune et al. (2016) that evaluated the ecoindustrial development of the Shanghai Chemical Industrial Park, China.

China has witnessed rapid development of the chemical industry and has become the largest chemical-producing country in the last decade, where more than 45% of the companies above a designated size have been clustered into a large number of chemical industrial parks. Greening the development of chemical industrial parks in China is crucial to the local environment and has long been a big challenge. Deploying IE strategies in a chemical industrial park will enhance both economic and environmental performances. This study reviewed the ecoindustrial development in the Shanghai Chemical Industrial Park (SCIP) and its performance. SCIP is a newly established industrial park producing petrochemicals and downstream fine chemicals and is considered a hallmark of the chemical industrial parks in China, with regard to ecoindustrial development from scratch and its good economic and environmental performances.

[c] https://www.usgs.gov/faqs/what-a-geographic-information-system-gis?qt-news_science_products=0#qt-news_science_products.

According to the authors, the key lessons drawn from SCIP are as follows:

(i) Its design and construction comply with a top-down, environmentally friendly pattern in line with IE strategy by integrating upstream and downstream chemicals, utilities and infrastructure, logistics, safety and environmental management, and public services holistically within the park.

(ii) SCIP created a vertically integrated chemical manufacturing network, from naphtha to polymers, to extend the value chain as much as possible via the geographical proximity of upstream-downstream linkage and economy of scale.

(iii) A delicately designed infrastructure and utility symbiotic network, logistic system, and 24-7 online safety and environmental management system sustained the food web efficiently.

Furthermore, the implementation of top-down ecoindustrial development planning (Fig. 16.3) and rich experience cumulated thereof in SCIP will be a

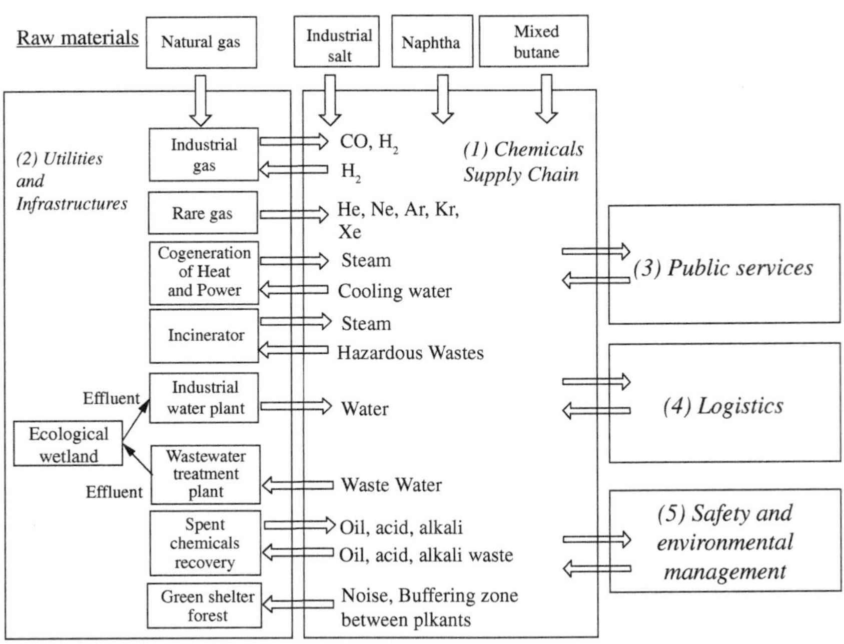

Fig. 16.3 The five-integration development strategy applied in SCIP toward industrial symbiosis. *(Credit: Yune, J.H., Tian, J., Liu, W., Chen, L., Descamps-Large, C., 2016. Greening Chinese chemical industrial park by implementing industrial ecology strategies: a case study. Resour. Conserv. Recycl. 112, 54–64. Reprinted with permission from Elsevier.)*

benchmark and enlightenment of key factors for facilitating green development of other chemical industrial parks in China.

16.4 Conclusions

As a scientific branch closely related to sustainability, IE proposes the development of production systems considering the mimetization of the dynamic nature for ecosystems that can promote positive environmental impacts on those processes and products based on RC. It is relevant to consider the partnership of IE with a circular economy and green chemistry (especially principle 7) to reach the condition of sustainability.

The application of IE approach in the chemical industry is a good example to be followed by several industry sectors and is not limited to chemical production.

References

Baldassare, B., Schepers, M., Bocken, N., Cuppen, E., Korevaar, G., Calabretta, G., 2019. Industrial symbiosis: towards a design process for eco-industrial clusters by integrating circular economy and industrial ecology perspectives. J. Clean. Prod. 216, 446–460.

European Commission, 2018. Industrial Symbiosis. https://ec.europa.eu/environment/europeangreencapital/wp-content/uploads/2018/05/Industrial_Symbiosis.pdf. accessed December 2021.

FISSAC Project, 2020. The Project. https://fissacproject.eu/en/project/. accessed December 2021.

Francis, C.G., 2003. The chemical industry from an industrial ecology perspective. In: Bourg, D., Erkman, S., Chirac, J.P. (Eds.), Perspectives on Industrial Ecology. Routledge, London.

Giurco, D., Prior, J., Boydell, S., 2014. Industrial ecology and carbon rights. J. Clean. Prod. 80, 211–223.

Yune, J.H., Tian, J., Liu, W., Chen, L., Descamps-Large, C., 2016. Greening Chinese chemical industrial park by implementing industrial ecology strategies: a case study. Resour. Conserv. Recycl. 112, 54–64.

CHAPTER 17

Remarks and conclusions of the sustainability aspects for renewable carbon

As discussed in Chapters 5 and 12 (Parts 1 and 2), remarks and conclusions are approaches used to facilitate book understanding and its applicability in real situations. Remarks cover the most representative statements described in the text, while conclusions critically summarize the conducted discussions. Then both can be used as a practical compilation, which take parts of relevant contents from each chapter. This didactic approach comprises:

✓ The reading of the text

✓ The understanding of its content

✓ The application of its fundamentals to the renewable carbon (RC) concept related to technological basis offered by Part 3.

17.1 Remarks

We saw in Chapter 13 that sustainability is a complex concept that needs to be addressed by the global society in the 21st century. In our technological world, there is a continuous search for the development of sustainable processes and products that can produce more valuable goods. The three pillars or components of sustainability are composed of a variety of internal indicators to be considered, such as: economic impacts (natural resources use, environmental management, and pollution prevention applied to air, water, land, and waste); social impacts (standard of living, education, community, equal opportunities); and economic impacts (profit, cost savings, economic growth, research and development). Furthermore, when we consider the processing of RC raw materials, there are some categories of specific sustainability indicators, mainly for environmental and social components, to achieve a reduced environmental footprint and increased societal value (United States Environmental Protection Agency, 2012).

Renewable Carbon
https://doi.org/10.1016/B978-0-323-99735-5.00017-7

We saw in Chapter 14 that to measure the economic, social, and environmental impacts of technologies taking advantage of RC, robust methodologies are needed to quantify and analyze their gains, especially environmental gains. Currently, lifecycle analysis (LCA) is considered the most efficient methodology in the study of environmental impacts and has been highly used in sectors of energy generation, polymer production, mining, transport, etc. The main characteristic of an LCA is the possibility of evaluating the environmental impacts of each component of the process, which allows for a more accurate quantification of the impacts of a given route and, consequently, to optimize the proposition of future goals and decision making. As an example of a tool used to measure and calculate, RenovaCalc was developed by the Brazilian Agricultural Research Corporation (2017) that quantifies the greenhouse gas (GHG) emissions over the lifecycle of biofuels, generating a performance index in $gCO_2eq\ MJ^{-1}$ of biofuel. Therefore, those that present production processes with less carbon emissions will have access to credits of carbon to be commercialized.

We saw in Chapter 15 that according to the International Energy Agency (2021), the importance of petrochemicals is continuing to grow, which is already a major component of the global energy system. Demand for plastics—the most familiar group of petrochemical products—has outpaced that of all other bulk materials (such as steel, aluminum, or cement) and has nearly doubled since 2000. Advanced economies, such as the United States and Europe, currently use up to 20 times as much plastic and up to 10 times as much fertilizer as developing economies, such as India and Indonesia, on a per capita basis. The growth in demand for petrochemical products means that petrochemicals are set to account for more than one-third of the growth in oil demand to 2030, and nearly half to 2050, ahead of trucks, aviation, and shipping. Petrochemicals are also poised to consume an additional 56 billion cubic meters of natural gas by 2030, equivalent to about half of Canada's total gas consumption today (International Energy Agency, 2021). Thus, RC raw materials are growing as alternative and sustainable feedstock for chemicals and materials derived from oil, going against principle 7 of green chemistry (use of renewable feedstock), especially in the case of biomass components and CO_2.

And we saw in Chapter 16 that we can consider industrial ecology (IE) a scientific branch closely related to sustainability because it proposes the development of production systems considering the mimetization of the dynamic of nature for ecosystems, which can promote positive environmental impacts on those processes and products based on RC sources discussed in Part 2. Moreover, IE is related to the concept of sustainability insofar as it involves closed production cycles, with reduction or elimination of dependence, for example, on nonrenewable energy sources. One of the pioneering cases on the application of the concept and principles of IE is the cluster of industries in the municipality of Kalundborg, Denmark.[a] There, a group of 9 large companies operates in a system in which more than 30 products are exchanged between them. For example, excess gas from an oil refinery process is treated to remove sulfur, which in turn is used to make sulfuric acid, which is used by two other industries in the cluster. In another case, the residue from the production of a pharmaceutical industry, in the form of biomass (a RC raw material), is used as fertilizer by farms in the region. Kalungborg's urban symbiosis has influenced parks around the world, from the United States to Japan.

Thus, sustainability for products and processes is a strategy that must be put in practice by considering and measuring their impacts associated with a holistic production system.

17.2 Conclusions

We concluded in Chapter 13 that sustainable development meets the needs of the present without compromising the ability of future generations to meet their own needs; reducing the negative impacts of modern society on the environment is paramount, as highlighted and proposed by the Sustainable Development Goals. To reach a sustainability condition, strategies such as circular economy, green chemistry, and carbon footprint calculation should be put in practice for RC source's industrial usage and valorization. In addition, it can be advantageous not only from the environmental point of view but also from the economic one as observed in carbon pricing efforts.

We concluded in Chapter 14 that LCA is highlighted as the main tool for sustainability measurements for several processes and products related to RC. Indeed, LCA is considered the most efficient methodology in the study of environmental impacts and has been highly used in sectors of energy generation, polymer production, mining, transport, biomass, etc. However, limitations must be considered to better estimate the projections resulting from the interpretation of the results, but they do not invalidate LCA. Thus, the application of calculation tools based on LCA promotes carbon mitigation by means of carbon quantification and pricing, as demonstrated for biofuels.

[a] http://www.symbiosis.dk/en/.

We concluded in Chapter 15 that there are no doubts that petrochemical products are everywhere and are integrated in modern society. They include plastics, fertilizers, packaging, clothing, digital devices, medical equipment, detergents, tires, and many others. They are also found in many parts of the modern energy system, including solar panels, wind turbine blades, batteries, thermal insulation for buildings, and electric vehicle parts. Indeed, oil demand as feedstock is growing on a global scale. However, the petrochemical industry is a source of pollution and contamination, requiring alternative feedstocks and technological strategies to reach sustainability as renewable building blocks.

Finally, we concluded in Chapter 16 that, as a scientific branch closely related to sustainability, IE proposes the development of production systems considering the mimetization of the dynamic nature for ecosystems, which can promote positive environmental impacts on those processes and products based on RC. It is relevant to consider the partnership of IE with a circular economy and green chemistry (especially principle 7) to reach sustainability. The application of the IE approach in the chemical industry is a good example to be followed by several industry sectors, not limited to chemical production.

Thus, as described in the three Parts of this book, science and technology should be integrated with sustainability to promote the RC approach considering modern technical and economic visions such as circular economy, bioeconomy, and green chemistry. This integration could boost wealth and well-being linked to lesser impacts on our environment.

References

Brazilian Agricultural Research Corporation, 2017. RenovaCalc. https://www.embrapa.br/busca-de-solucoes-tecnologicas/-/produto-servico/4355/renovacalc-#20-#20-ferramenta-para-comprovacao-do-desempenho-ambiental-de-usinas-de-biocombustiveis. accessed December 2021.

International Energy Agency, 2021. The Future of Petrochemicals—Towards a More Sustainable Chemical Industry. https://www.iea.org/reports/the-future-of-petrochemicals. accessed December 2021.

United States Environmental Protection Agency, 2012. A Framework for Sustainability Indicators at EPA. EPA. https://www.epa.gov/sites/production/files/2014-10/documents/framework-for-sustainability-indicators-at-epa.pdf. (accessed December 2021).

Further reading

United Nations, 2021. Sustainable Development Knowledge Platform. https://sustainabledevelopment.un.org/. accessed December 2021.

Index

Note: Page numbers followed by *f* indicate figures, *t* indicate tables, and *s* indicate schemes.

D

E

L

M

N

O

P

R

S